THE LAST DARK SKY

How We Are Lighting Up the Night

A Natural and Unnatural History of Night, from Fire to Starlink

Douglas Donovan

IBSN: 979-8-2562-7509-9

Introduction — Look Up

It is midnight in Las Vegas, and the desert refuses to be dark.

Atop the Luxor hotel, a xenon beam engineered to be the single brightest artificial light on the planet rises out of a black glass pyramid and drills a column of white fire straight into the atmosphere. It is visible, on a clear night, from an airplane hundreds of miles away. It is visible from space. It was not built to illuminate a street, guide a ship, or warn a plane away from a tower. It exists, more or less, because it is possible — a monument to the idea that light is cheap, darkness is free for the taking, and no one in particular owns the sky above a casino roof.

The beam does something its designers did not advertise. It gathers insects by the hundreds of thousands, drawn upward out of the desert floor into a column of light that must look, to a moth's simple eye, like an object more compelling than the moon ever was. Where insects gather, bats follow, wheeling through the beam all night to feed on a food source that would not exist without the fixture above them. Sky, beam, insects, bats — an entire miniature ecosystem, built by accident, out of waste light.

A few hundred miles away, across the border in Chile, or a continent away on a mountaintop in the Canary Islands, an astronomer is doing something that sounds almost absurd when placed next to the Luxor beam: waiting, patiently, hour after hour, to capture photons that have been traveling since before the Earth existed. Some of that light left its source eleven or twelve billion years ago. It crossed the expanding universe, missed every star, dodged every dust cloud, and arrived, at last, at a mirror built at staggering expense specifically to catch it before it dissipates into the cosmic background forever.

Those two scenes — a beam thrown upward for no reason at all, and a telescope straining to catch light that has traveled for eons — are not really separate stories. They are the same story, told from opposite ends. This is a book about that story: how humanity, over the course of only a few centuries, engineered its way out of the natural night, and what it is now discovering, almost too late, about the cost of that victory.

For nearly the entire history of our species, the night sky was not a special occasion. It was furniture. It was where farmers found the seasons, where sailors found their way home, where every mythology on Earth built its gods and monsters, where shepherds and priests and emperors alike watched the same wheeling stars and argued about what they meant. A person born in almost any century before the twentieth could walk outside on a clear night and see the Milky Way arch overhead as an actual structure — not a rumor, not a photograph, but a visible river of light, dense with the combined glow of hundreds of billions of stars in our own galaxy, seen edge-on from within.

That experience, so ordinary for so long, is now the exception. Roughly a third of humanity — and the overwhelming majority of people in North America and Europe — can no longer see the Milky Way from where they live. Children are being born and raised entirely inside a version of night that has never, in the four-billion-year history of Earth's darkness cycle, existed before: a night with a floor of light glowing up from below, bright enough to erase the dimmer 90 percent of what the naked eye could otherwise perceive.

This book tells that story in two parts, because it turns out there are two distinct assaults on the night sky, arriving from two different directions, on two very different timelines.

The first is old news, or at least it should be. It began with fire and oil lamps and gaslit streets, accelerated explosively with Thomas Edison's incandescent bulb and the electrification of cities, and has been compounding for more than a century as humanity built brighter, taller, and more numerous light sources with almost no consideration of what happens to the light that misses its intended target and simply escapes upward. This is terrestrial light pollution: skyglow, the orange-white dome that now sits over every city on Earth, dense enough in some places to be visible from three hundred miles away. It is a story of engineering, economics, and a stubborn cultural assumption — largely unexamined until recently — that more light is always safer, always better, always progress.

The second assault is new, and it is the reason this book exists now rather than twenty years ago. For all of human history, if terrestrial light pollution got bad enough, there was an answer: leave. Drive far enough into the desert, climb high enough up a mountain, sail far enough out to sea, and eventually you would find real darkness again, the kind unspoiled by any city's glow. That option is what allowed astronomers to keep building bigger and better telescopes on remote mountaintops through the twentieth century, chasing darkness to the ends of the Earth. It is what dark-sky tourists still chase today in the Atacama Desert of

Chile, one of the driest, darkest, most astronomically prized landscapes on the planet.

But escape assumed the problem stayed on the ground. It no longer does. Since 2019, a single company has launched more satellites into low Earth orbit than existed in all of human history combined before that year, with tens of thousands more approved or proposed by competing operators. These satellites are bright enough, in the right conditions, to be seen with the naked eye crossing the sky in bright trains, and they are numerous enough to leave visible streaks across the long-exposure images that professional telescopes depend on. Terrestrial light pollution can, in principle, be escaped by distance. Orbital light pollution cannot. It does not matter how remote your observatory is, how empty the desert around it, how many international treaties protect the darkness of the land below — the satellites pass overhead regardless.

That is the sharpest and strangest turn this story takes, and it deserves to be said plainly, without exaggeration in either direction: we are not yet at the point where ground-based astronomy is finished. Large telescopes still have enormous advantages that space telescopes cannot easily replicate, and clever engineering — coordination software, satellite-darkening coatings, careful orbital design — is already reducing some of the damage. But we are, credibly and measurably, approaching a point where the darkness of the sky itself may become a limiting resource for what ground-based astronomy can accomplish — a resource as finite and contested as clean water or open spectrum, being spent down by parties who mostly are not thinking about astronomy at all.

There is something almost perfectly ironic in the shape of this problem, and the irony is worth sitting with, because it will recur throughout this book in different forms. We invented electric light, and we used it to illuminate the night so thoroughly that we lost the ability to see the stars from our own backyards. In response, we built telescopes of extraordinary size and sensitivity, on the darkest mountaintops we could find, specifically to see the stars we could no longer see at home. And now, in the middle of doing that, we have begun filling the sky itself — the medium the telescopes look through — with thousands of our own bright, reflective, humming machines.

We are, in other words, simultaneously reaching farther into the universe than any humans in history and making it harder than ever to simply look up and see it. That is not a minor engineering footnote. It is a question about what kind of civilization we intend to be, and it does not have an obvious answer yet.

This book will take you through both halves of that story: down to Earth first, into the history and physics and biology of terrestrial light pollution — where it came from, what it costs ecosystems and human health, and where the last dark places on the planet still survive. Then up, into orbit, into the astonishing and still-unfolding story of satellite megaconstellations and what they mean for the future of looking up at all.

Along the way we will return, more than once, to a simple visual device: the same patch of night sky, viewed first from a truly dark location — the kind of place where the Milky Way is not merely visible but overwhelming — and then from progressively brighter vantage points, until the same sky, over a large city, shows perhaps a few dozen of the thousands of stars that were there all along, merely drowned out. Watching the sky disappear step by step, rather than all at once, is the clearest way to understand what has actually been lost, and how recently, and how fast.

We will also come back, more than once, to Las Vegas, and to that absurd, magnificent, wasteful beam atop the Luxor — because it is, in miniature, the whole argument of this book. It is proof that we can throw as much light as we want into the sky, for no better reason than that we can. And it is a standing invitation to ask the question this book keeps returning to, in a dozen different forms, on the ground and in orbit alike: is the night sky something we own, to use up however we like — or something we merely inherited, and are obligated, eventually, to hand on?

Go outside tonight, if you can, somewhere reasonably dark, and look up. What you see — or fail to see — is the subject of this book.

How this book is organized

The chapters that follow are organized in two large movements, plus this introduction and a closing epilogue that bookend them.

The first movement — Chapters 1 through 10 — stays almost entirely on the ground. It begins with the night sky as it existed for the whole of human history before artificial light, then follows the technological history that dismantled that sky piece by piece: fire and oil lamps, gaslight, the electric bulb, the neon sign, the floodlit skyscraper, the interstate highway, and the casino strip. It introduces the Bortle scale as a tool for measuring exactly how much of the sky has been lost, and shows, with as much precision as the available data allows, how much of the world's population has already lost the Milky Way entirely. It then turns to consequences — first ecological, then human — before closing with an account of the last genuinely dark places remaining on Earth, and the observatories built to exploit them.

The second movement — Chapters 11 through 14 — leaves the ground behind and goes into orbit. It tells the much more recent, much faster-moving story of satellite megaconstellations: how they came to exist, what they actually do to astronomical images, and where current deployment trends might lead over the next decade or two. It asks, without pretending to answer with false precision, what all of this means for the long-term future of looking at the universe from the surface of the Earth.

Chapter 15 turns from diagnosis to response, cataloging what has actually worked — shielded fixtures, warmer color temperatures, dark-sky ordinances, satellite-operator mitigation commitments — and what remains unresolved. The epilogue returns, one final time, to the beam of light in the Nevada desert with which this introduction began, and to the question that beam has represented throughout: whether the night sky is ours to spend, or ours to protect.

Readers primarily interested in the history of artificial lighting may find the first movement most rewarding; readers who came to this book because of Starlink and its competitors may be tempted to skip ahead to Chapter 11. I would gently discourage that shortcut. The orbital story only makes full sense against the terrestrial one — the same assumptions, the same unexamined logic of "more light because we can," the same absence of anyone with clear authority to say no, recur in both halves of this story, separated by a century and a half but connected by a single throughline. Reading the book in order is the fastest way to see that throughline clearly.

A note on sources and certainty

This book draws on a combination of well-established historical record, peer-reviewed scientific literature, industry and regulatory filings, and, for the most speculative material in its later chapters, careful extrapolation clearly flagged as such. Where the underlying evidence is strong — the history of electrification, the physics of skyglow, the documented ecological impacts on sea turtles and migratory birds — this book states its claims plainly. Where the evidence is genuinely more preliminary or contested — some of the human health associations discussed in Chapter 8, the longer-range satellite population projections discussed in Chapter 13 — this book has tried, throughout, to say so explicitly rather than borrow false confidence from adjacent, better-established findings. Readers are encouraged to treat that distinction as seriously as the book itself has tried to.

A word, finally, on scope. This book is written primarily from the vantage point of the industrialized, heavily light-polluted regions of North America and Europe, and much of its historical narrative reflects that vantage point directly, simply because that is where the technologies and institutions this book traces were first developed and most thoroughly documented. Its underlying argument, however, is not regional. Every city on every populated continent is, to varying degrees and on varying timelines, moving through some version of the same story this book tells, and the orbital half of that story, by its very nature, touches every latitude on Earth without exception.

Chapter 1 — The Night We Inherited

Before there was a single artificial light anywhere on the planet, night was not empty. It was arguably the most information-dense experience available to a human being — a slowly wheeling instrument panel that told you the season, the hour, the latitude, the date of the coming harvest, and, if you believed the priests and philosophers who studied it most closely, the mood of the gods themselves. To understand what has been lost to artificial light, it helps to first understand, in some detail, what was actually there.

What a truly dark sky looks like

Modern readers, almost all of whom have only ever known a sky diluted by artificial light to one degree or another, tend to underestimate what an unpolluted night sky actually delivers. It is not merely "more stars." It is a fundamentally different visual experience.

Under a genuinely pristine sky — the kind now surviving only in a small number of protected deserts, remote islands, and high-altitude sites — the unaided human eye can perceive somewhere between two and five thousand individual stars at any one time, roughly half the sky's total visible population, the other half being below the horizon. That is merely the beginning. The Milky Way itself, our galaxy seen edge-on from within, is not a faint smudge in a truly dark sky. It is a structured, textured band of light running from horizon to horizon, bright enough to cast faint shadows, with visible dark lanes of interstellar dust threading through its bright core, dense star clouds in

Sagittarius and Scorpius standing out as genuinely brighter regions, and a clear asymmetry between the crowded galactic center and the sparser anti-center direction. People seeing it for the first time in their adult lives — a common experience now, on trips to dark-sky parks — frequently do not recognize it. They ask what the cloud is.

Beyond the Milky Way, a dark sky reveals the zodiacal light: a faint cone of light rising from the horizon along the ecliptic before dawn or after dusk, caused by sunlight scattering off interplanetary dust in the plane of the solar system. It is subtle — easily washed out by even modest light pollution or a bright moon — but visible to a patient, dark-adapted eye almost anywhere on Earth's surface, if the sky itself is dark enough. Ancient observers in genuinely dark locations would have seen it as a matter of routine; today, most people who have heard of the zodiacal light at all have heard of it as a curiosity, not something they have witnessed themselves.

There is also the simple matter of stellar color. Under sufficient darkness and with enough dark adaptation, the unaided eye begins to discern color in the brighter stars: the orange-red of Betelgeuse and Antares, the blue-white of Rigel and Vega, the golden tint of Arcturus. Light pollution and, ironically, excessive ambient brightness from any source — including the moon — wash this subtlety away long before it reduces the sheer count of visible stars.

The instrument nobody had to build

None of this was decorative to the people who lived under it. Before mechanical clocks, before printed calendars, before GPS, the night sky was functioning civil infrastructure — arguably the single most widely shared piece of infrastructure in human history, available identically to a farmer in the Nile valley, a herder in Mongolia, and a fisherman in Polynesia, requiring no institution, no fuel, and no maintenance beyond clear weather.

Agricultural societies across essentially every continent independently discovered that the appearance and disappearance of particular stars at dawn or dusk — their heliacal risings and settings — tracked the solar year far more reliably than watching the sun itself, whose position along the horizon shifts too gradually to notice day to day. The star Sirius rising just before the sun, for instance, told ancient Egyptian farmers that the Nile's annual flood was imminent, a piece of information valuable enough that the Egyptian calendar was built around it. Pleiades risings and settings served an equivalent function across cultures from ancient Greece to Polynesia to the Andes, marking planting and harvest windows with a precision that no other available instrument could match.

Maritime navigation depended even more directly on the stars, and did so for millennia before any alternative existed. Polynesian navigators crossed thousands of miles of open Pacific Ocean using a memorized "star compass" — a mental catalog of which stars rose and set at which points on the horizon, cross-checked against ocean swell patterns and bird behavior, that allowed voyages of astonishing precision without instruments of any kind. European navigation depended on similar principles for centuries, refined but not fundamentally replaced until the chronometer and, much later, satellite positioning. In both traditions, and dozens of others besides, competent use of the night sky was not a specialized science reserved for elites; it was baseline literacy, as essential and as commonly held as knowing how to read weather.

Timekeeping at night — a genuine problem before artificial illumination, since sundials are useless after dark — was likewise solved by the sky itself. The apparent rotation of the celestial pole stars (whichever star happened to occupy that role at a given epoch, since precession slowly changes the answer over thousands of years) allowed the estimation of the hour by the rotation angle of the surrounding constellations, a technique sophisticated enough that specialized instruments — nocturnals — were built in the medieval and early modern periods specifically to read the time from the position of the Big Dipper relative to the pole star.

Myth, religion, and the sky as shared text

Because the night sky was universally visible and endlessly patterned, it became, independently, in culture after culture, the medium onto which humanity projected its deepest stories. This was not a coincidence of imagination; it was a practical consequence of the sky being the largest, most detailed, most permanent visual surface any pre-industrial society had continuous access to.

The constellations themselves are one of the clearest examples of humans imposing narrative structure onto an essentially arbitrary pattern of points, and then treating that narrative as functionally real. Greek mythology mapped hunters, queens, and monsters onto the sky and used those figures to explain everything from the changing seasons to moral lessons about hubris. Babylonian astronomers, centuries earlier, had already developed one of the first systematic zodiacs, tracking planetary movement against a backdrop of fixed constellations for both religious and what we would now call proto-scientific purposes — the two were not clearly separated. Chinese astronomy developed an entirely independent constellation system organized around an Emperor's court analogy, with the pole star as the unmoving Emperor around whom the rest of the sky's "officials" revolved — a striking example of how the same physical sky can support entirely different cultural readings depending on what a society needs it to mean.

Religious cosmology across cultures routinely placed the night sky at or near the center of belief systems. The Egyptian sky goddess Nut was depicted arching over the earth god Geb, her body forming the vault of heaven itself, with the sun traveling through her each day and being reborn each dawn. Mesoamerican civilizations, notably the Maya, built entire calendrical and religious systems around precise astronomical observation, tracking Venus's cycle with an accuracy that still impresses modern astronomers, and aligning major temple structures to solstice and equinox sunrise points with a precision that required generations of careful record-keeping. Indigenous Australian traditions, among the oldest continuously practiced cultures on Earth, encode detailed astronomical knowledge — including, in some traditions, awareness of variable stars and the general structure of the Milky Way as a river or path of ancestor spirits — inside oral traditions many thousands of years old.

None of these traditions treated the night sky as scenery. They treated it as text: something to be read, interpreted, and argued over, the way later societies would argue over scripture or law. The disappearance of that shared text — not the disappearance of astronomy as a specialized profession, but the disappearance of naked-eye access to the night sky as a piece of common cultural property — is one of the quieter losses this book will return to, particularly in the chapter on the human and psychological costs of light pollution.

The physics of an unlit night

It is worth pausing on why the pre-electric night sky was as dark as it was, because the physical baseline matters for understanding how thoroughly it has since been altered.

In the complete absence of artificial light, the night sky's brightness is set by a small number of well-understood natural sources: starlight itself, scattered and re-scattered faintly through the atmosphere; the zodiacal light already described; a very faint, uniform atmospheric airglow caused by chemical reactions high in the upper atmosphere, invisible to casual observation but measurable with instruments; and, on nights when it is present, moonlight, which can by itself increase sky brightness by a factor of hundreds compared to a moonless dark sky. Astronomers use the term "natural sky background" to describe this baseline, and it forms the reference point against which all subsequent light pollution is measured — the darkest a sky can possibly get, absent artificial sources, at a given location and lunar phase.

This baseline was, for the entire span of human prehistory and the vast majority of human history, essentially the only sky condition that existed anywhere on Earth after sunset, apart from the transient light of fires. It is the sky that every human culture, in every location, developed its navigation, agriculture, timekeeping, mythology, and cosmology in direct relationship with. It is also, not coincidentally, close to the sky condition astronomers now travel to remote deserts and mountaintops to recover, at enormous expense, because it no longer exists within any reasonable driving distance of the majority of the world's population.

The scale of what changed, and how quickly

The rest of this book is, in one sense, an extended answer to a single question: how did we get from that baseline — universal, free, culturally load-bearing — to a present in which the same experience requires either extraordinary remoteness or expensive travel to preserve it?

The honest and somewhat unsettling answer is that almost all of the change happened extremely recently, on the timescale of human history. Fire has been used for illumination for perhaps a million years, but firelight is weak, localized, and does not meaningfully brighten the sky beyond its immediate vicinity — a campfire does not create skyglow visible from the next valley. Oil lamps, candles, and even gaslight, which we will examine in the next chapter, brightened streets and homes but still left the sky itself essentially untouched at any distance; a person a few miles outside an eighteenth-century gaslit city still saw close to a natural night sky.

The transformation that actually erased the stars for most of humanity is barely more than a century old, and the more severe phase of it — bright, ubiquitous, poorly shielded electric lighting spreading across entire metropolitan regions and beyond — is well within living memory for people alive today. Someone born in a mid-sized American or European city in the 1950s could plausibly have seen the Milky Way from their childhood backyard; a child born in the same city today, statistically, cannot, and may never have reason to leave the artificially lit region surrounding that city long enough to see it anywhere.

That is the pace this book is actually describing: not a slow geological drift, but a civilization-scale environmental transformation compressed into roughly a hundred and fifty years, with the steepest part of the curve occurring in the most recent fifty. Understanding what came before it — the sky this chapter has described, inherited essentially unchanged from the birth of the species until shockingly recently — is the necessary baseline for everything that follows. The next chapter turns to how the dismantling began: not with electricity, but with the first tentative attempts to hold back the dark using fire, oil, and gas.

The Bortle ladder, previewed

Later chapters will introduce a formal instrument — the Bortle Dark-Sky Scale, developed in 2001 — for describing exactly how much artificial light has intruded on a given night sky. It is worth previewing the two extremes of that scale here, in the context of this chapter's description of the pre-electric night, because the contrast is the clearest possible illustration of what this book is actually about.

At the darkest end of that later scale, Class 1, the description is essentially identical to what this chapter has already described as the ordinary, unremarkable condition of every pre-electric night on Earth: the Milky Way overhead as a bright, structured band, capable of casting faint shadows; the zodiacal light visible as a matter of routine; thousands of individual stars visible simultaneously. At the opposite end, Class 9 — the sky over a dense urban core today — the Milky Way is entirely invisible, only a handful of the brightest stars and planets remain visible at all, and the sky itself glows with a distinct artificial color, orange or yellow-white depending on the local lighting technology. Every human being who lived before the twentieth century experienced something close to Class 1 as their baseline, ordinary night. A rapidly growing share of humanity today experiences something close to Class 8 or 9 as their baseline, ordinary night, and would need to travel a considerable distance, at real expense of time and money, to experience anything resembling what their own ancestors saw for free, simply by walking outside.

This is the single central fact this book is built around, and it will recur constantly across the chapters that follow: a form of experience that was, for effectively the entire history of the human species, universal, free, and unremarkable, has become, within less than two human lifetimes, rare, effortful, and increasingly the province of specialized travel and expensive scientific instrumentation.

Star lore as functional technology, not mere decoration

It is worth dwelling further on a point raised earlier in this chapter: the mythological and religious traditions built around the night sky across virtually every human culture were not simply decorative flourishes layered onto an otherwise unrelated body of practical astronomical knowledge. In many documented cases, the mythological framework and the practical, functional use of the sky were the same body of knowledge, encoded in a form that made it memorable, transmissible across generations without writing, and resistant to error.

Consider Polynesian wayfinding in more detail, since it is among the best-documented and most sophisticated examples of star-based navigation as functional technology. Traditional Polynesian navigators memorized what is often called a star compass: a mental division of the horizon into distinct houses or sectors, each associated with the rising or setting point of specific stars or star groups at specific times of year. A navigator crossing open ocean — voyages that, in the case of the settlement of Hawaii, New Zealand, and Easter Island, covered thousands of kilometers of featureless Pacific with no landmarks whatsoever — would track their heading continuously against this memorized star compass, cross-checking against wave patterns, cloud formations associated with distant islands, and the flight patterns of specific seabird species known to range a certain distance from land and no farther. This system was sophisticated enough, and accurate enough, that experimental voyages in the twentieth century using replica canoes and traditionally trained navigators successfully retraced historically documented Pacific crossings using nothing but these methods, definitively demonstrating that the techniques described in oral tradition were not myth or exaggeration but a genuinely functional navigational technology, fully capable of the voyages historically attributed to it.

European celestial navigation developed along a parallel but independent track, relying more heavily, in its pre-instrument form, on the pole star's fixed position and the rotation of circumpolar constellations around it to establish latitude and heading, techniques that predate and, for centuries, coexisted alongside the magnetic compass rather than being immediately replaced by it, since a compass alone cannot determine latitude the way careful observation of stellar altitude can.

Calendars built from the sky

Agricultural calendars represent a third major category of functional, sky-derived technology, alongside navigation and mythology, and deserve slightly more attention than the earlier, brief mention of Egyptian and Andean examples. The ancient Egyptian civil calendar, one of the oldest continuously documented calendrical systems in history, was explicitly built around the heliacal rising of Sirius — the first morning the star became visible again after a period of being hidden by the sun's glare — because that specific stellar event reliably preceded the annual Nile flood by a consistent, short interval, making it a far more dependable agricultural signal than watching the river itself, whose flood timing could not be predicted in advance without some independent, earlier signal.

The Maya civilization of Mesoamerica developed an even more elaborate example: a calendrical system that tracked, with an accuracy still considered impressive by modern researchers working from surviving astronomical records, the synodic cycle of Venus, integrating that cycle into both agricultural planning and an elaborate religious and political calendar used to time everything from planting to warfare to royal ceremonies. Structures at major Maya sites are oriented with a precision that requires sustained observation across many years to achieve, aligning specific temple doorways or platform edges to the sunrise or sunset point on solstices, equinoxes, or other astronomically significant dates — architecture, in other words, functioning as a kind of permanent astronomical instrument, embedding a civilization's accumulated sky knowledge directly into its built environment in a form that would remain useful for the same purpose for many generations after the original observers who designed it had died.

What this chapter is not arguing

It is worth being precise, before moving on to the technological history that follows, about what this chapter has not claimed. It has not claimed that pre-electric life was, on balance, preferable to modern life, or that electrification and its benefits — genuinely enormous benefits, in public health, safety, productivity, and quality of life — were a mistake. Nothing in this book argues for abandoning electric lighting. The argument, developed across the chapters that follow, is considerably narrower and more specific: that the particular way electrification was pursued — with essentially no consideration, for the first century and more of its history, of where the light that missed its intended target actually went — produced a genuinely unnecessary and still largely uncorrected cost, layered on top of electrification's real and substantial benefits, and that correcting that specific design flaw, as later chapters will show, does not require giving up any of those benefits at all.

That correction begins with understanding, in careful historical detail, exactly how the original design flaw came about. The next chapter turns to that history.

The moon as a competing, natural light pollutant

It is worth noting, before leaving this chapter's discussion of the pre-electric sky, that natural darkness was never perfectly uniform even before artificial light existed. The moon itself, waxing and waning across its roughly 29.5-day cycle, functioned as a natural, entirely predictable source of periodic sky brightening, washing out the fainter stars, the zodiacal light, and much of the Milky Way's finer structure for several nights around each full moon, before receding again as the cycle continued toward new moon. Pre-industrial societies were, as a matter of routine practical knowledge, keenly aware of this cycle and planned nighttime activities requiring darkness — certain religious observances, specific hunting and fishing practices timed to lunar phase, and, for many cultures, the calendrical systems discussed earlier in this chapter — around it explicitly.

This natural variability matters for this book's argument because it establishes that the pre-electric sky was never a single, unchanging condition, but a predictable, cyclically varying one, in which even the darkest, most star-filled nights occurred on a knowable schedule, alternating with brighter, moonlit nights during which the more dramatic array of stars, though still present, was considerably less visually striking. Artificial light pollution differs from this natural lunar cycle in a crucial respect: it does not cycle or recede. A full moon eventually wanes; a floodlit parking lot, a lit billboard, or an orbiting satellite constellation does not, absent deliberate human choice, ever grant the sky back its darkest nights on any comparable schedule. This distinction — a natural, temporary, cyclical brightening our ancestors planned around, versus an artificial, sustained, non-cyclical brightening that simply accumulates indefinitely absent deliberate intervention — is worth keeping in view throughout the historical chapters that follow.

Anthropological breadth: a few additional traditions

The examples already discussed in this chapter — Egyptian, Polynesian, Babylonian, Chinese, Maya, and Indigenous Australian sky traditions — could be supplemented almost indefinitely, and it is worth naming a handful of additional traditions briefly, simply to underscore how genuinely universal this pattern of sky-based cultural elaboration actually was. Norse mythology built an elaborate cosmology around the aurora and the visible stars, with specific constellations tied to particular sagas and figures. West African griot oral traditions in several documented cultures incorporated star-based seasonal markers into agricultural planning cycles passed down through generations of trained oral historians. Inuit traditions in the Arctic, operating under sky conditions dramatically different from equatorial or temperate cultures — including extended periods of both continuous winter darkness and continuous summer daylight — developed correspondingly distinctive astronomical knowledge systems adapted to those specific extreme conditions, including detailed knowledge of specific stars' positions used for orientation during the long polar night, when the sun itself was unavailable as a directional reference for extended periods.

This breadth matters because it forecloses any suggestion that the cultural significance of the night sky described in this chapter was somehow a parochial feature of a small number of well-documented "major" civilizations. It appears to have been close to a universal feature of human cultural development, arising independently and repeatedly wherever human societies existed, precisely because the underlying resource — a clear, dark night sky — was, until the period this book describes, universally and freely available to virtually every human community on the planet, regardless of that community's size, technological sophistication, or geographic isolation from other cultures.

The eclipse as a reminder

One further, more recent piece of evidence for everything this chapter has argued is worth mentioning: total solar eclipses, which briefly and artificially recreate something close to a natural night sky in the middle of the day, along the narrow path of totality. Observers experiencing a total eclipse for the first time consistently report the sudden appearance of bright stars and planets in a sky that was, moments before, an ordinary bright blue daytime sky, and this sudden, dramatic reintroduction of a star-filled sky to people who may never otherwise have seen one is frequently described, in both scientific and popular writing about eclipse tourism, as one of the most viscerally affecting parts of the entire eclipse experience — a brief, involuntary demonstration, granted by celestial mechanics rather than travel to a dark-sky site, of exactly the experience the rest of this book argues has been engineered away from most people's ordinary lives.

Sailors' logs as an overlooked scientific record

Ships' logs from the age of sail, maintained meticulously for navigational and legal purposes across many centuries of maritime trade, occasionally recorded incidental celestial observations - meteor showers, comet sightings, unusually vivid auroral displays - that modern researchers have mined as an unplanned but genuinely valuable historical dataset documenting the pre-electric night sky as experienced by trained, literate observers across a wide range of latitudes and historical periods, offering an additional, independent line of evidence corroborating the general picture of near-universal dark-sky access this chapter has described through more conventional historical and anthropological sources.

What survives in language itself

One further, subtler piece of evidence for how thoroughly the night sky once permeated ordinary human life survives in language itself: a striking number of everyday words and idioms across many languages derive from celestial phenomena - words for months tied directly to lunar cycles, expressions invoking stars, luck, and fate that trace back to older astrological and cosmological frameworks. This linguistic residue persists today largely as fossilized idiom, detached from any living, daily observational practice, in one more quiet testament to how recently, and how completely, direct experiential contact with the phenomena that originally generated this language has receded from ordinary life.

Chapter 2 — The First Light

Long before Thomas Edison and the incandescent bulb, humans had already begun, tentatively and then more confidently, to push back against the dark. The story of that early push is less dramatic than the electrification that would follow, but it matters, because it establishes a pattern that recurs throughout this book: every new lighting technology arrives justified by safety and commerce, spreads faster than anyone anticipated, and leaves behind consequences nobody planned for.

Fire, the first and longest-lived technology

For the overwhelming majority of the roughly one million years humans have controlled fire, its relationship to the night sky was essentially neutral. A campfire, a torch, a hearth — these produce enough light to see by at a range of a few meters to perhaps a few dozen meters, and essentially nothing at the scale of a sky. Even the largest pre-industrial settlements, lit entirely by wood fires, oil lamps, and later candles, did not brighten their local sky enough to be measurably different from the surrounding countryside on a clear night. This is worth stating plainly because it is easy to assume, incorrectly, that light pollution is simply what happens whenever humans gather and use light; in fact it required specific technological thresholds — chiefly, cheap and plentiful fuel, and later, the electric grid — before artificial light became powerful and ubiquitous enough to compete with starlight.

Oil lamps burning animal fat, olive oil, or later whale oil represented the first meaningful improvement in controllable, indoor and outdoor light over open fire, offering a steadier flame and, with the addition of a wick, a more directional and controllable light source. Whale oil in particular became a major global industry by the eighteenth and nineteenth centuries, driving an international hunting economy specifically to fuel lamps in wealthier cities and ships, until petroleum-derived kerosene displaced it in the latter half of the nineteenth century — cheaper, more available, and arguably one of the few instances in industrial history where a fossil fuel substitution reduced pressure on a wild animal population, since whale oil demand collapsed rapidly once kerosene became available.

Candles, meanwhile, evolved from tallow — rendered animal fat, cheap but smoky and foul-smelling — to beeswax, prized for its cleaner burn but expensive enough to remain a luxury item for churches and the wealthy for centuries, to, eventually, industrially produced paraffin and stearin candles in the nineteenth century that brought clean-burning candlelight within reach of ordinary households for the first time. None of these technologies, however widely adopted, meaningfully altered the sky. They lit rooms, streets immediately around a lamp post, and small courtyards. The night beyond the reach of any given flame remained dark in the way it always had been.

The invention of the street as a lit space

The genuinely consequential shift in this period was not a new fuel but a new idea: that public streets, as a matter of civic policy, should be illuminated after dark, for reasons of safety and order rather than any individual's convenience. This idea has a surprisingly specific and traceable history.

Paris is usually credited as an early and influential adopter, with a 1667 royal ordinance under Louis XIV mandating lanterns be hung at regular intervals across the city, following a wave of concern about nighttime crime and disorder in a rapidly growing capital. London followed with its own street-lighting requirements in the following decades, and by the eighteenth century most major European capitals had some form of mandated, if inconsistent, public lighting — typically oil lamps maintained by contracted lamplighters who made nightly rounds to light, extinguish, and refuel each fixture, a labor-intensive system that limited how extensively any city could realistically light itself.

The justification offered for this new civic expense was almost always framed in terms of public safety: lit streets, the argument went, deterred crime, reduced accidents, and made cities more governable after dark, when unlit streets had traditionally been effectively lawless in ways that made policing difficult and travel dangerous. This safety-and-order justification is worth noting carefully, because it is essentially the same justification offered for every subsequent expansion of artificial lighting for the next three and a half centuries, right through to contemporary debates over LED streetlight brightness — and it is a justification that, as later chapters will show, has repeatedly outrun the actual evidence supporting it, even as it has proven durable enough to survive that mismatch.

Gaslight: the first true infrastructure

The technology that changed the scale of the problem, well before Edison, was gas lighting. Coal gas — produced by heating coal in the absence of oxygen and piping the resulting flammable gas through a network of underground mains — was first demonstrated for street lighting in London in the first years of the nineteenth century and had spread to most major European and American cities within a few decades. This was a genuinely different kind of technology from oil lamps or candles: it required, for the first time, a centralized production facility (a gasworks), a distribution network (underground pipes), and a permanent, continuously burning supply rather than fuel carried and refilled lamp by lamp. In other words, gaslight was the first lighting technology that was, in the modern sense, infrastructure — a public utility with the scaling logic of infrastructure, which is to say a logic that tends only to expand.

Gaslit cities of the mid-to-late nineteenth century were, by the standards of everything that had come before them, extraordinarily bright. Contemporary accounts from the period describe visitors to gaslit London or Paris being almost disoriented by how thoroughly the nighttime streets were illuminated compared to anything in their home regions, and gaslight quickly became not merely a safety measure but a matter of civic pride and international prestige — cities competed, quite explicitly, over how brilliantly and extensively they were lit, treating illumination as a visible marker of modernity and wealth in exactly the way that skyscraper height or airport size would later function for twentieth-century cities.

This competitive dynamic matters because it introduces, for the first time in this book's story, a motive for lighting that has nothing to do with safety or utility at all: light as spectacle, light as status, light as an end in itself. Gaslit boulevards, illuminated public monuments, and gas-lit exhibition halls at world's fairs in this period were early, low-resolution previews of exactly the impulse that would later produce neon Las Vegas and the Luxor beam — the discovery that a city, or an institution, or eventually a private company, could use conspicuous brightness as a form of communication independent of any practical lighting need.

The first complaints

It is worth noting — because it undercuts any assumption that concern over light pollution is a purely modern, post-industrial phenomenon — that gaslight generated its own contemporary critics almost immediately, on grounds that will sound familiar. Some objected on aesthetic grounds, arguing that gaslight's harsh, flickering quality was inferior to the softer light of oil lamps or candles it was replacing. Others objected on safety grounds specific to the new technology itself: gas leaks, explosions, and asphyxiation were genuine and recurring hazards of early gas lighting systems, and public anxiety about gasworks explosions was a real feature of nineteenth-century urban life. A smaller number of observers, particularly among astronomers and naturalists, noted with concern that the brightening of city skies was beginning to interfere with naked-eye stargazing in ways that had simply not been possible before — early, isolated instances of what would later be systematically documented as skyglow, though the term itself did not yet exist and no one had yet thought to measure the phenomenon scientifically.

These early complaints did not slow the expansion of gas lighting in any meaningful way, and this too establishes a pattern: from the earliest gaslit streets of the 1810s through the LED conversions of the 2010s, concerns about the unintended costs of new lighting technology have almost never been sufficient, on their own, to change the trajectory of adoption. Adoption has been driven by cost, convenience, and competitive civic pressure; concerns about side effects have generally been addressed, when they are addressed at all, only after the technology is already ubiquitous.

The stage set for electricity

By the final decades of the nineteenth century, then, the essential shape of the modern lighting problem was already visible, even though the sky itself had not yet been transformed nearly as dramatically as it soon would be. Cities had committed to the principle that public space should be lit after dark. A genuine infrastructure — the gasworks and pipe network — existed to support that commitment at scale, and had already demonstrated that lighting infrastructure, once built, tends to expand rather than contract. Lighting had already become entangled with civic pride, competition, and spectacle, not merely utility. And a handful of early observers had already noticed, and worried about, the effect of brightening skies on the visibility of the stars.

What was missing was a light source powerful, cheap, and controllable enough to take all of these existing tendencies and multiply them by an order of magnitude. That source arrived in the final quarter of the nineteenth century, and its story — electrification, and the astonishingly rapid transformation of the world's cities that followed — is the subject of the next chapter.

The economics of early lighting technology

It is worth pausing on the specific economics of pre-electric lighting fuels, because cost dictated access far more directly than it would in the electric era that followed, and the pattern of who could afford light, and how much of it, shaped both the pace and the character of the transition described in this chapter.

Tallow candles, rendered from animal fat, were cheap enough to be within reach of most households in preindustrial Europe and colonial America, but they burned with a notoriously smoky, foul-smelling flame and required frequent replacement, making sustained bright illumination of an entire home for an entire evening a genuine and recurring expense even for households that could technically afford the candles themselves. Beeswax candles burned far more cleanly and brightly but cost dramatically more to produce, since beeswax is a comparatively scarce byproduct of honey production rather than a readily available agricultural surplus, and their use was largely restricted to churches, wealthy households, and special occasions for centuries.

Whale oil occupies a particularly interesting place in this economic history because it briefly became, in the eighteenth and early nineteenth centuries, one of the most valuable traded commodities in the Atlantic economy, driving a global hunting industry — centered substantially in New England ports such as Nantucket and New Bedford — specifically to supply lamp oil for both domestic lighting and, critically, street lighting in growing cities on both sides of the Atlantic. The price of whale oil rose considerably over the first half of the nineteenth century as accessible whale populations were depleted by the expanding hunting fleet, a dynamic that in retrospect functioned as an early, unplanned experiment in resource depletion driving up the cost of an essential lighting fuel faster than any deliberate conservation policy could have. The discovery and commercial refinement of kerosene from petroleum in the 1850s arrived at almost exactly the right moment to solve this looming supply problem, offering a cheaper, more abundant illuminant that rapidly displaced whale oil for both lamp and, within a few decades, street-lighting use, a substitution that is sometimes credited, not entirely facetiously, with saving several whale species from a trajectory toward commercial extinction that the whale-oil lighting industry alone might otherwise have driven them toward.

Lamplighters and the labor of early public lighting

The practical operation of gaslight and earlier oil-lamp street lighting depended on an occupation that has almost entirely disappeared from living memory but that represented, for the better part of two centuries, a significant urban workforce: the lamplighter, employed by a city or a private lighting contractor to make nightly rounds lighting each fixture at dusk and, in many jurisdictions, extinguishing them again near dawn, refilling oil reservoirs or checking gas valves along the route, and reporting broken or malfunctioning fixtures for repair. A mid-sized nineteenth-century city might employ dozens of lamplighters working overlapping routes each evening, and the profession had its own informal customs, tools (a long pole with a wick-lighting flame at one end and a hook or valve-key at the other, depending on the fuel type), and even, in some cities, its own recognized associations or guilds.

This labor-intensive operational model imposed a real practical ceiling on how extensively any given city could realistically light itself, since every additional fixture added to a lighting network meant additional nightly labor to service it, a constraint that dissolved almost overnight with electrification, since electric fixtures, once wired into a grid, required no nightly manual lighting or extinguishing at all and could in principle be left burning indefinitely with no ongoing labor cost beyond the electricity itself. This shift — from a technology whose expansion was throttled by available labor to one whose expansion was throttled only by capital investment and electricity generation capacity — is one of the underappreciated structural reasons electrification, once it began, expanded so much faster and so much more completely than gaslight ever had, and it is a pattern worth keeping in mind heading into the next chapter's discussion of just how quickly that expansion actually occurred.

Early aesthetic and safety objections, in more detail

The brief mention earlier in this chapter of contemporary objections to gaslight deserves slightly more elaboration, both because these objections established a template for lighting-technology criticism that recurs throughout the rest of this book's history, and because they illustrate how differently early critics framed their concerns compared to the environmental and astronomical framing this book itself adopts.

Aesthetic critics of gaslight in the nineteenth century frequently focused on the specific quality of the light itself: gas flames of the period tended to flicker and vary in brightness depending on gas pressure fluctuations within the distribution network, producing a less steady, less pleasant illumination than a well-tended oil lamp or candle, and early gas fixtures produced noticeable soot and combustion byproducts that discolored ceilings, curtains, and artwork in enclosed indoor spaces, a complaint that became a genuine driver of continued candle and oil-lamp use in wealthier indoor settings even as gaslight took over public streets and commercial premises. Safety critics, meanwhile, focused overwhelmingly on the risk of leaks and explosions inherent in a citywide network of pressurized, combustible gas mains and household connections, a risk that was not merely theoretical: gasworks explosions and residential gas fires were a recurring feature of nineteenth-century urban incident reporting, and insurance industry data from the period reflects meaningfully elevated premiums for gas-lit commercial and residential structures compared to their oil-lit or candle-lit counterparts.

What is almost entirely absent from the surviving record of nineteenth-century lighting criticism, by contrast, is any sustained public concern about the effect of increasingly bright cities on the visibility of the night sky, beyond the isolated astronomer and naturalist complaints already noted. This absence is itself instructive: it suggests that the environmental and astronomical costs of artificial lighting, however real, remained essentially invisible to the general public and to policymakers throughout this entire early period, not because those costs did not exist, but because no one had yet developed the observational tools, the institutional interest, or the vocabulary necessary to notice, measure, and articulate them as a problem worth public attention — a gap that would not begin to close in any serious way until well into the twentieth century, as the chapters following the next will describe.

Gas street lighting's spread beyond Europe's capitals

While this chapter has focused primarily on London and Paris as the emblematic early adopters of gaslight, it is worth noting that the technology spread with considerable speed to a much wider range of cities across Europe and North America over the following decades, following a broadly similar pattern in each case: initial adoption in a city's wealthiest commercial and civic core, followed by gradual, multi-decade expansion into residential neighborhoods, with poorer and more peripheral areas generally receiving gas lighting last, if at all, before the technology itself began to be superseded by electricity in the closing decades of the nineteenth century.

American cities followed this same general trajectory with a lag of roughly a decade or two behind the leading European examples: Baltimore is generally credited as the first American city to establish a public gas lighting utility, in 1816, with major cities including New York, Boston, and Philadelphia following over the subsequent two decades. This transatlantic spread illustrates a pattern that will recur throughout this book's history: a new lighting technology, once demonstrated successfully in one or two leading cities, tends to spread rapidly to other cities through a combination of direct technical emulation, competitive civic pressure (no aspiring major city wished to be conspicuously less modern than its rivals), and the straightforward commercial interest of gas equipment manufacturers and utility operators in expanding their business to new markets as quickly as the available capital and technical expertise allowed.

Domestic lighting versus public lighting: two parallel stories

It is worth distinguishing more clearly, at this point in the book's historical narrative, between two related but distinct threads this chapter has woven together: the story of public street and civic lighting, funded and organized by municipal authorities and driven by the safety and civic-pride motivations discussed above, and the separate story of domestic and commercial indoor lighting, driven primarily by individual household and business purchasing decisions rather than municipal policy.

These two threads developed on somewhat different timelines and were driven by somewhat different considerations. Public street lighting expansion, as this chapter has described, was substantially a policy-driven, civic-competition-driven phenomenon, shaped by the safety and prestige considerations already discussed. Domestic and commercial indoor lighting improvement, by contrast, was driven considerably more directly by simple household and business economics: as candle, oil, and eventually gas lighting became cheaper and more broadly accessible across the socioeconomic spectrum through the nineteenth century, ordinary households and small businesses adopted brighter and more convenient lighting simply because it improved their own immediate quality of life and productivity, largely independent of any broader civic or policy consideration.

Both threads, however, contributed to the same eventual outcome relevant to this book's argument: a cumulative, citywide increase in total artificial light output, some of it publicly mandated and organized, much of it privately and individually chosen, all of it contributing, in aggregate, to the same skyglow phenomenon discussed in later chapters, regardless of whether any individual light source was a municipally installed street lamp or a private household's parlor lighting. This dual-threaded historical pattern — civic policy and individual private choice both contributing to the same aggregate environmental outcome — is a structural feature of the light pollution problem that persists directly through to the present day, and that any comprehensive policy response, as Chapter 15 discusses, must account for both dimensions simultaneously rather than addressing only the more easily regulated public-lighting thread in isolation.

Street lighting's legal foundations

The specific legal and administrative mechanisms nineteenth-century cities used to fund and mandate street lighting are worth a brief note, since they established administrative patterns that persist, in modified form, in the municipal lighting ordinances discussed in Chapter 15. Most cities funded early public lighting through a combination of general property taxation and, in a number of documented cases, special lighting assessments levied specifically on property fronting a newly lit street, on the theory that adjoining property owners received a direct, quantifiable benefit from the improvement and should bear a proportionate share of its cost, a funding logic that closely parallels the special-assessment financing still used for many local infrastructure improvements today.

Insurance records as an unlikely historical archive

Historians studying the pace and geography of early gaslight and electrification have found insurance company records to be an unexpectedly rich, underused archival source, since insurers maintained detailed property surveys noting the presence and type of lighting installed in insured commercial buildings, for underwriting purposes entirely unrelated to any interest in lighting history as such. These records allow researchers to reconstruct, with a level of geographic and temporal precision rarely available from newspaper accounts or civic records alone, exactly which specific blocks and buildings within a given city had converted to gas or electric lighting by a given year, providing some of the best available quantitative evidence for the block-by-block, rather than merely city-by-city, pace of the transition this chapter describes.

The candle tax and unintended consequences

Britain's candle tax, levied intermittently between the seventeenth and nineteenth centuries and eventually repealed in 1831 partly in response to public health arguments that it discouraged adequate home lighting among poorer households, offers an early and instructive example of government policy directly shaping lighting access along economic lines, a dynamic that recurs, in modified form, in the equity concerns raised later in this book regarding uneven access to both artificial lighting improvements and, eventually, dark-sky recreational amenities.

Fire codes and the birth of lighting-adjacent regulation

Municipal fire codes, developed and strengthened considerably through the nineteenth century in response to a series of devastating urban fires, incidentally shaped early lighting practice by imposing restrictions on flame-based lighting in certain building types and densities, a regulatory precedent that, while motivated purely by fire-safety concerns rather than any interest in sky darkness, established the basic principle that lighting practice was a legitimate subject of municipal regulation at all, a principle the dark-sky ordinances discussed in Chapter 15 would eventually draw on far more directly.

Chapter 3 — Edison Changes the Night

If gaslight was the dress rehearsal, electric light was the performance that never ended. Within roughly two human generations — from the 1880s to the 1930s — cities across the industrialized world converted from gas and oil lighting to electric lighting, and in doing so, changed the relationship between humanity and the night sky more dramatically, and more permanently, than any development in the preceding several thousand years of civilization.

The bulb and the system

Thomas Edison's incandescent light bulb, commercially demonstrated in 1879 and rapidly refined over the following years, is popularly remembered as the invention that changed nighttime lighting. That framing is not wrong, but it is incomplete in a way that matters for this book's argument. The bulb itself was only one component of a much larger system Edison and his contemporaries had to build simultaneously: centralized power generation, a distribution grid to carry electricity from generating stations to individual buildings, metering to make the whole system commercially viable, and a manufacturing base capable of producing bulbs, wiring, and fixtures cheaply and in enormous volume. Edison's Pearl Street Station in lower Manhattan, which began supplying electricity to a few dozen customers in 1882, was less a lighting product than the prototype of an entire industry — one that would, within a few decades, become one of the largest and most capital-intensive infrastructure systems in the industrialized world.

This system-level character of electrification is central to understanding why the resulting increase in artificial light was so much larger, and so much harder to reverse, than anything gaslight had produced. Gas networks, however extensive, were fundamentally limited by the physical difficulty and cost of piping combustible gas over long distances and into every building. Electrical grids, once built, could in principle extend indefinitely, generation capacity could be added incrementally, and — critically — electric light was safer, cleaner, and required none of the ongoing maintenance of a fuel-burning system. There was no lamplighter to schedule, no wick to trim, no risk of asphyxiation. Electric light was, in almost every practical sense, a strictly superior product to everything that had come before it, which is precisely why its adoption proceeded so much faster and so much more completely than gaslight's had.

The World's Fairs as demonstration and spectacle

If any single set of events crystallized, for a global public, the idea that electric light represented civilizational progress itself, it was the great World's Fairs and international expositions of the late nineteenth and early twentieth centuries. The 1893 World's Columbian Exposition in Chicago — the famous "White City" — deployed tens of thousands of incandescent bulbs to illuminate an entire fairground of neoclassical buildings each night, an experience so far beyond anything most of the millions of visitors had previously witnessed that contemporary accounts describe audible gasps from crowds encountering the illuminated fairground for the first time. The 1901 Pan-American Exposition in Buffalo pushed the same idea further with its "Tower of Light," a purpose-built structure whose sole function was to be covered in as many bulbs as technically feasible and lit up as a nightly spectacle.

These fairs were not merely lighting demonstrations; they were deliberate propaganda for electrification as an idea, financed in substantial part by electrical equipment manufacturers and utility companies who had an obvious commercial interest in creating public appetite for a technology that was, at the time, still expensive, unfamiliar, and viewed with some suspicion by the general public. The strategy worked. Visitors returned home from these expositions as advocates for electrification in their own towns and cities, having witnessed firsthand a vision of nighttime brightness that had, until that visit, existed nowhere on Earth. This is an important and underappreciated mechanism in the history of light pollution: the demand for ever-brighter nighttime lighting was not simply a natural response to need, but was actively manufactured, at critical junctures, by industries with a direct financial stake in the outcome — a dynamic that, as later chapters will show, recurs in different forms right through to contemporary LED marketing and satellite-internet business models.

The astonishing speed of urban electrification

Once the basic commercial and technical case for electric lighting was established, its spread through major cities was remarkably fast by the standards of any prior infrastructure transition. New York's electric streetlights expanded from a handful in the early 1880s to citywide coverage within roughly two decades. London, Paris, Berlin, and other major European capitals underwent broadly similar transitions on a similar timescale, typically converting their central districts first and their outer neighborhoods over the following twenty to thirty years, with gas lighting persisting in some peripheral and rural areas well into the twentieth century even as it vanished from city centers.

What made this transition different in kind, not just degree, from the gaslight era that preceded it, was the sheer output of the new fixtures combined with a near-total absence of any consideration for where the light actually went. Early electric streetlights, and the incandescent fixtures rapidly adopted for storefronts, office buildings, and eventually private homes, were essentially unshielded — bare bulbs or simple reflectors that threw light in every direction with no attempt to direct it downward, toward the street or sidewalk that actually needed illumination, rather than upward and sideways into the sky and neighboring buildings. This lack of shielding, adopted purely because it was cheaper and simpler to manufacture unshielded fixtures, is directly responsible for a substantial fraction of the skyglow problem that persists in cities to this day; light that serves no functional purpose whatsoever, because it never illuminates any surface a person actually needs to see, nonetheless escapes upward and scatters off atmospheric particles to create the glowing dome now visible over every substantial city on the planet. The physics of this scattering — how airborne dust, water vapor, and pollution particles catch and redistribute upward-directed light back down toward observers many miles away — would not be systematically studied and quantified for the better part of another century, but the light itself began accumulating in the sky from the very first decade of large-scale electrification.

Brightness as an unquestioned good

Perhaps the most consequential feature of this period, from the perspective of a book about light pollution, is what did not happen: at no point during the initial decades-long rollout of electric lighting across the world's cities was there any serious public discussion of whether ever-increasing brightness was actually desirable, or whether there might be a point of diminishing returns, or negative returns, beyond which more light produced more harm than benefit. The assumption that brightness equaled safety, brightness equaled progress, and brightness equaled prosperity was treated as self-evident, not as a hypothesis requiring evidence.

This assumption was reinforced from multiple directions simultaneously. Municipal governments, competing with rival cities for reputation and investment, treated extensive and modern lighting as a marker of civic success, exactly as they had with gaslight a generation earlier, only now with a technology capable of vastly greater intensity. Electrical utilities, whose entire business model depended on selling as much electricity as possible, had an obvious commercial incentive to encourage maximal lighting use and none whatsoever to encourage restraint or efficient shielding, since unshielded fixtures that wasted light upward consumed exactly as much electricity, and generated exactly as much revenue, as efficient ones that directed all their output usefully downward. Insurance companies and early urban-safety advocates promoted brighter lighting as crime prevention, a claim that, as later chapters will discuss, has always been more intuitively appealing than empirically well-supported, but which proved durable and difficult to dislodge once established as conventional wisdom.

The cumulative effect of these converging incentives was a lighting culture, established in the first few decades of the electric era and essentially unchanged in its basic assumptions for the following century, in which more light was axiomatically better, the cost of wasted upward-scattered light was invisible and therefore irrelevant to any decision-maker, and no institution anywhere had either the incentive or the mandate to ask whether the resulting transformation of the night sky was a cost worth counting.

The sky begins to disappear

It is difficult to reconstruct, with precision, exactly how fast any given city's sky darkened — sky-brightness measurement is a twentieth-century scientific development, discussed in detail in Chapter 5, and simply did not exist as a discipline during the crucial early decades of electrification. But indirect evidence is abundant and consistent. Amateur astronomy societies in major cities, some dating to the nineteenth century, left records of members relocating their observing activities progressively farther from city centers across the early twentieth century, chasing darkness that had previously been available within city limits. Newspaper astronomy columns from the 1900s through the 1930s periodically noted, with a kind of resigned commentary, that this or that celestial event — a modest meteor shower, a faint comet — would be difficult to observe from within city limits due to "the glare of the streetlights," language that appears with increasing frequency as electrification proceeded.

By the middle of the twentieth century, the basic shape of the problem this book addresses was already established, even though almost no one at the time would have used the term "light pollution" to describe it, and the more severe compounding effects of postwar suburban expansion, highway lighting, and illuminated advertising — the subject of the next chapter — had not yet occurred. Cities were substantially brighter than they had been fifty years earlier. That brightness had been achieved with essentially no attention to efficiency, direction, or necessity. And the sky above the world's major population centers had already begun the transformation that would, within another two generations, place the Milky Way permanently out of reach for the majority of people living in or near them.

The next chapter follows that transformation into its most extravagant and unrestrained phase: the mid-twentieth-century city of neon, floodlit skyscrapers, and illuminated highways, culminating in a desert city that would eventually build a beam of light powerful enough to be seen from space, and ask, implicitly, what light is actually for.

Edison, Westinghouse, and the war of the currents

No account of early electrification is complete without at least a brief treatment of the "war of the currents" — the commercial and technical rivalry between Thomas Edison's direct current (DC) electrical distribution systems and the alternating current (AC) systems promoted by George Westinghouse and developed substantially by Nikola Tesla, whose induction motor and polyphase AC system provided the crucial technical foundation Westinghouse's company commercialized. This rivalry, fought out over roughly a decade beginning in the mid-1880s through a combination of technical argument, public demonstration, and a notably aggressive public-relations campaign by Edison's side emphasizing (with real but selectively presented justification) the electrocution risks of high-voltage AC distribution, ultimately resolved in AC's favor for large-scale power distribution, for a straightforward and durable technical reason: alternating current can be transformed to very high voltages for efficient long-distance transmission and then stepped back down to safer voltages near the point of use, a capability direct current systems of the period could not easily replicate, making AC distribution dramatically more practical for lighting and powering entire cities, let alone the larger regional grids that would eventually connect them.

This resolution mattered enormously for the pace of the transformation described in this chapter, because AC's transmission efficiency meant that a relatively small number of large, centralized generating stations could economically supply electricity across a much wider geographic area than the localized DC generating stations of Edison's original Pearl Street model, allowing electrification to spread outward from initial urban cores into surrounding neighborhoods and eventually entire metropolitan regions far more quickly and cheaply than a purely DC-based system could have achieved. The specific choice of AC over DC as the dominant distribution standard, in other words, is not a mere technical footnote; it is one of the underlying enabling conditions for the rapid, near-total urban electrification this chapter describes, and by extension, one of the underlying enabling conditions for the scale of artificial light growth this entire book is concerned with.

The rural electrification gap, and what it reveals

It is worth noting that the electrification story told so far in this chapter is substantially an urban story, and the gap between urban and rural electrification timelines in many countries is itself informative. In the United States specifically, rural areas lagged urban centers by several additional decades in most cases, with a large fraction of American farms remaining unconnected to any electrical grid well into the 1930s, a gap significant enough that the federal government established the Rural Electrification Administration in 1935 specifically to extend electrical infrastructure into areas private utilities had judged insufficiently profitable to serve on their own, given the lower population density and correspondingly higher per-customer infrastructure cost of rural distribution compared to dense urban service territory.

This urban-rural gap matters for this book's argument for a specific reason: it means that the sky-brightening effects described in this chapter, severe as they were in and around major cities by the early twentieth century, remained considerably more limited in the extensive rural and small-town areas that had not yet been electrified, preserving reasonably dark skies within a much shorter distance of any given city than would be the case by the mid-to-late twentieth century, once rural electrification programs, and the postwar suburban and highway development discussed in the next chapter, closed that gap almost completely across most of the industrialized world. The buffer of genuine rural darkness that still existed, in most regions, through the first third of the twentieth century, is part of why amateur astronomy remained broadly practical from a much wider range of locations during this period than would be the case only a few decades later.

Early efficiency arguments, largely ignored

It would be inaccurate to suggest that no one, during the early decades of electrification, ever raised efficiency or waste as a consideration in fixture design; some lighting engineers and municipal officials of the period did note, in technical trade publications and occasional municipal reports, that unshielded fixtures wasted a considerable fraction of their output in directions that served no illuminating purpose, and a handful of more thoughtful early street-lighting designs did incorporate basic reflector shields specifically to direct more light downward and reduce glare for pedestrians and, incidentally, reduce upward waste as well.

These efficiency-minded design choices, however, were adopted inconsistently, driven primarily by a desire to reduce glare and improve downward illumination for practical, ground-level reasons — better visibility for pedestrians and vehicles, reduced light spilling uncomfortably into nearby windows — rather than any articulated concern about sky brightness or astronomical impact specifically, which remained, as noted at the end of the previous chapter, essentially outside the frame of public and professional concern during this period. The result was a fixture landscape that varied considerably in efficiency and shielding quality from city to city and even fixture to fixture within a single city, but that trended, in aggregate, toward the same basic pattern described throughout this chapter: enormous, historically unprecedented growth in total artificial light output, with only incidental and inconsistent attention to where the light that missed its target actually went, setting the stage for the even more extravagant, spectacle-driven lighting culture the next chapter describes.

The incandescent bulb's own technical evolution

It is worth noting that the incandescent bulb itself continued to improve substantially in the decades following Edison's original commercial demonstration, in ways that further accelerated the lighting expansion this chapter describes. Early carbon-filament bulbs, while a genuine and commercially viable improvement over any prior lighting technology, were relatively inefficient by later standards, converting only a small fraction of their electrical input into visible light and the remainder into heat, and had comparatively short operational lifespans requiring frequent replacement. The introduction of tungsten filaments in the first decade of the twentieth century, considerably more durable and efficient than the original carbon filaments, produced brighter, longer-lasting, and more economical bulbs, further accelerating adoption by improving the basic cost-benefit calculation facing any household, business, or municipality considering conversion from gas or continued reliance on older bulb technology.

This pattern of continuous incremental technical improvement driving continued adoption and expansion is one this book will observe recurring at multiple later points in its history: LED lighting technology, discussed at length in Chapter 15, followed a broadly similar trajectory in the more recent past, with each successive generation of LED fixtures offering improved efficiency and reduced cost, driving accelerating adoption in a pattern directly comparable to the tungsten-filament bulb's role in accelerating electrification a century earlier — a reminder that the underlying economic and technical dynamics driving lighting expansion have remained remarkably consistent across more than a century of otherwise quite different specific technologies.

Electrification's international variation

While this chapter has focused primarily on American and, to a lesser extent, European electrification, it is worth noting briefly that the pace and pattern of electrification varied considerably across other regions of the world, generally correlating closely with each region's broader industrial development timeline. Japan's major cities electrified on a broadly comparable timeline to leading European and American cities, reflecting the country's rapid industrialization in the late nineteenth and early twentieth centuries. Colonial territories under European administration generally saw electrification concentrated disproportionately in administrative and commercial centers serving colonial governance and extraction interests, often with limited extension into surrounding areas serving primarily local populations, a pattern whose legacy in some regions persisted well into the postcolonial period. Large parts of what would later be described as the developing world did not achieve widespread urban electrification until considerably later in the twentieth century, and in some rural regions, reliable grid electricity remains incomplete even today, a genuinely important qualification to keep in mind when this book's later chapters discuss "global" light pollution trends, since the specific historical timeline this chapter has traced applies most directly and completely to the already-industrialized regions where the bulk of currently measured light pollution originates, even as rapid electrification and urbanization in additional regions, discussed in Chapter 6, has become an increasingly significant contributor to global light pollution growth in more recent decades.

Municipal power ownership debates

The rapid electrification described in this chapter also fueled a significant political debate in many American and European cities over whether electrical generation and distribution should be owned and operated by private utility companies or by municipal governments directly, a debate that shaped the pace and character of lighting expansion in ways worth briefly noting. Advocates of municipal ownership argued that a service as essential as street lighting and general electrification should not be subject to private profit motives that might favor underserved poorer neighborhoods or extract excessive rates from captive customers; private utility advocates countered that private capital and competitive pressure would deliver electrification more efficiently and with fewer of the corruption risks associated with municipal contracting. Both models were adopted in different cities, and the resulting patchwork of ownership structures across different municipalities is part of why the specific pace and completeness of electrification, and the resulting lighting standards adopted, varied as much as it did from city to city even within the same country and the same historical period.

A brief digression on the vocabulary of "progress"

The rhetoric surrounding electrification during this period consistently invoked the word "progress" in ways worth pausing on, since that same rhetorical move recurs, largely unchanged, in later marketing for LED conversion, smart-city lighting initiatives, and satellite broadband deployment alike. In each case, a new lighting or connectivity technology is framed as an unambiguous civilizational advance, its costs, where acknowledged at all, treated as minor engineering footnotes rather than genuine tradeoffs deserving public deliberation. This book has no interest in denying that electrification, LED efficiency gains, and satellite broadband connectivity are each, in their own right, genuine advances; the argument, consistent from this chapter through the book's close, is simply that "progress" has rarely been treated as a claim requiring evidence about its full range of consequences, upward light included, rather than a self-evident conclusion.

Advertising and the birth of the illuminated brand

The final years of the nineteenth century also saw the first deliberate use of electric lighting specifically for brand advertising rather than general illumination, with early electric signs spelling out company names appearing on prominent buildings in major commercial districts, a direct technological and conceptual precursor to the neon-driven advertising culture examined at length in the next chapter, and an early, concrete illustration of electric light's rapid expansion beyond pure utility into deliberate commercial spectacle.

Household electrification's slower, uneven curve

Even as street and commercial electrification proceeded at the rapid pace this chapter has described, full household electrification, particularly for lower-income urban residents, generally lagged by a decade or more, meaning that for a meaningful transitional period, brightly electrified public streets coexisted with still gas-lit or candle-lit private homes, a stratified pattern of adoption that meant the sky-brightening effects this chapter describes accumulated somewhat unevenly across a city's full built environment even during the core decades of the electrification era.

Chapter 4 — The City That Never Sleeps

By the middle of the twentieth century, the electrified city had acquired a new vocabulary of light, one built not around illuminating streets for safety but around illuminating buildings, signs, and entire skylines for effect. This chapter follows that vocabulary from its origins in early neon signage through the illuminated skyscraper and the interstate highway, arriving finally at the city that took the logic of unrestrained nighttime brightness further than any other place on Earth: Las Vegas, Nevada.

Neon and the birth of light as language

Neon lighting, first demonstrated publicly in Paris in 1910 and commercialized through the 1920s, introduced something genuinely new to the history of artificial light: light as text and image, rather than merely illumination. A neon tube, bent into the shape of letters or pictures and filled with an inert gas that glows a specific color when electrified, could spell out a business's name in a way ordinary bulbs could not easily replicate, and could do so in colors — the reds, blues, and greens produced by different gas fills and glass tints — that had no precedent in earlier lighting technology.

American cities adopted neon signage with tremendous speed through the 1920s and 1930s, and by the postwar period, neon-lit commercial strips had become one of the defining visual features of American urban and roadside culture: Times Square, whose electrified advertising had already impressed visitors in the early bulb era, became substantially more visually dense and chromatically varied once neon and, later, animated electric signage were layered on top of the older illuminated billboards. Commercial strips along major roads in cities across the country adopted the same logic at a smaller scale, motel and diner signage in particular becoming an entire vernacular art form built around competing for a passing driver's attention through sheer visual intensity.

The significant shift here, for the purposes of this book, is not merely aesthetic. Neon and its successors introduced a category of lighting whose entire purpose was to attract attention rather than to illuminate a functional space — light designed to be looked at directly, rather than light designed to help someone see something else. This distinction matters because it removes any pretense of efficiency-driven design from the equation. A streetlight that wastes light upward is failing at its stated purpose, however common that failure may be. An advertising sign that blazes as brightly as physically possible, in every direction, including straight up into the sky, is succeeding at its stated purpose. There is no engineering fix for advertising-driven light pollution in the way there eventually would be, imperfectly, for street lighting; the waste is not a byproduct of poor design, it is frequently close to the entire point.

The floodlit skyscraper

A parallel development occurred at the scale of entire buildings. Architectural floodlighting — using powerful exterior lights to illuminate the facade of a building from outside, making it visible and often dramatic against the night sky — became a significant feature of American urban design beginning in the interwar period and accelerating dramatically after the Second World War. The Empire State Building's exterior lighting, periodically changed to mark holidays and events beginning in the 1930s and formalized into a regular programmed lighting scheme by the 1970s, is perhaps the best-known example, but the practice became widespread across major American and, eventually, global cities: illuminating a skyscraper's exterior became a routine expectation of prestige commercial architecture, not an unusual flourish reserved for a handful of landmark buildings.

The lighting engineering involved in floodlighting a large building inherently directs enormous amounts of light outward and upward — a facade illuminated from ground-mounted floodlights is, by definition, being lit from below, at an angle, meaning a substantial fraction of the light involved necessarily continues past the building into the sky above it rather than terminating usefully on the illuminated surface. Unlike street lighting, where shielding technology exists to direct light downward with reasonable efficiency, illuminating a vertical facade from an oblique angle has no equivalent efficient solution; some amount of sky-directed waste is close to structurally unavoidable in the basic physics of the technique, whatever fixture quality is used.

Highways and the standardization of brightness

The postwar decades also saw the construction of the interstate highway system in the United States and equivalent high-capacity road networks elsewhere, and with them, a new and enormous category of nighttime lighting: continuous or semi-continuous roadway illumination across thousands of miles of highway, interchange, and parking infrastructure. Highway lighting standards developed during this period were set, understandably, with a strong safety-first orientation, but the specific brightness levels, fixture spacing, and shielding requirements adopted were based on relatively limited research, often erred toward brighter rather than dimmer out of an abundance of caution, and were essentially never revisited or reduced once established, even as later research began to suggest that uniformity of lighting and glare control mattered considerably more for driver safety than sheer brightness.

The scale of highway and parking-lot lighting matters enormously for the overall light pollution problem because of its sheer geographic extent. A single downtown's illuminated skyscrapers, however bright, occupy a relatively compact area. A national highway network, together with the parking lots, service plazas, and commercial development that grew up alongside it, extends artificial lighting continuously across enormous swaths of countryside that would otherwise remain genuinely dark, effectively erasing the buffer zone between cities that had, in earlier eras, still offered reasonably dark skies within a short distance of any urban center. Suburban sprawl, expanding rapidly across the same postwar decades and closely tied to highway construction, compounded this effect further, converting what had been dark agricultural or undeveloped land at a city's periphery into continuously lit residential and commercial development, pushing the boundary of meaningfully dark sky farther and farther from any given urban population.

Las Vegas: the logical endpoint

If the preceding developments — neon signage as attention-seeking rather than functional light, architectural floodlighting with inherent upward waste, and the standardization of bright highway and parking lighting — represent the general trajectory of twentieth-century American nighttime lighting, Las Vegas represents that trajectory pursued to its most extreme and unapologetic conclusion, in a single city, by design.

Las Vegas's development as a resort and casino destination beginning in earnest after the Second World War was built, from very early on, around the deliberate use of extravagant, round-the-clock lighting as the city's core marketing identity. Unlike other American cities, where illuminated advertising and floodlit architecture developed alongside other civic priorities, Las Vegas's casino-resort economy had a direct and immediate commercial interest in maximizing visible brightness and spectacle, since the entire business model depended on drawing visitors from a considerable geographic radius specifically to see and experience that spectacle. Neon signage on Las Vegas's early casino strip through the 1940s and 1950s quickly became more elaborate, more animated, and more visually dominant than equivalent commercial signage anywhere else in the country, competing casinos essentially engaged in a decades-long arms race of ever more elaborate illuminated facades, each new resort attempting to visually overwhelm its predecessors.

By the later twentieth century, this competitive dynamic had produced a cityscape genuinely unlike anywhere else on the planet: a downtown core and adjoining resort corridor whose combined light output, viewed from orbit at night, is distinctly and immediately recognizable against the surrounding dark desert, forming one of the most visually striking artificial light signatures visible from space anywhere in North America. The desert setting is not incidental to this effect; because Las Vegas sits in an otherwise sparsely populated and naturally arid, clear-skied region with minimal ambient light for a considerable radius in every direction, the contrast between the city's artificial brightness and the surrounding natural darkness is unusually stark, making the city's light output appear even more extreme by comparison than an equivalent amount of lighting would in a more continuously developed region.

The Luxor beam: light with no function beyond itself

The single most vivid expression of this citywide logic is the Luxor Sky Beam, installed atop the Luxor Hotel's distinctive black glass pyramid when the resort opened in 1993. Engineered specifically to be the most powerful beam of light produced by a single fixture anywhere on Earth, the installation uses a cluster of high-intensity xenon lamps, combined through an arrangement of mirrors into a single vertical column, to project light directly upward with a total output frequently described, in various technical accounts over the years, as equivalent to tens of thousands of ordinary incandescent bulbs concentrated into one beam. On clear nights the beam is visible from aircraft at very considerable distances and has been reported as visible, under the right atmospheric conditions, from as far away as Los Angeles — several hundred kilometers distant.

What makes the Luxor beam significant for this book is not simply its intensity, considerable as that is, but its total absence of any illuminating function whatsoever. It does not light a street. It does not aid navigation, though early aviation-era lighthouse-style beacons occasionally served that function in other contexts. It illuminates no facade, guides no pedestrian, deters no crime in any location a person could plausibly occupy. It exists, as closely as any large-scale lighting installation on Earth comes to a pure case study, purely as spectacle: light produced because a resort wanted a signature visual element more distinctive than any competitor's, with the practical side effects of that light — its impact on the surrounding desert sky, its attraction of enormous numbers of insects drawn upward into the beam's column, and the bats that in turn gather nightly to feed on those insects, an entire small artificial food web assembled by accident around a light fixture with no ecological purpose at all — treated as, at most, a curiosity rather than a cost to be weighed against the beam's marketing value.

The Luxor beam will recur at several points later in this book, particularly in the chapter on ecological consequences, where its insect-and-bat food web offers an unusually clean, self-contained illustration of how artificial light restructures animal behavior even in a location with no other confounding factors. For now, it stands as the logical endpoint of the trajectory this chapter has traced: a full century of urban lighting increasingly organized around visibility, spectacle, and competitive brightness rather than function, arriving finally at a single beam of light thrown into the sky for no better reason than that the technology to do so existed and no institution had ever been positioned, or motivated, to ask whether it should.

The next chapter turns from this historical narrative to a more analytical question: given a century and a half of largely unrestrained lighting growth, how do we actually measure how dark — or how bright — a given sky has become? The answer is the Bortle scale, and the broader science of quantifying skyglow that has developed around it.

The crime-and-lighting assumption, examined

This chapter has already noted, in passing, that brighter lighting has long been promoted as a crime-deterrence measure, an assumption inherited directly from the civic street-lighting justifications of the seventeenth and eighteenth centuries described in Chapter 2. It is worth examining that assumption in somewhat more detail here, since it recurs as a justification for lighting expansion throughout the entire period this chapter covers, and since more rigorous evaluation of the claim did not really begin until relatively recently.

The intuitive case for brighter lighting reducing crime is straightforward: better visibility should make criminal activity more likely to be observed and identified, and darkness has long been associated, in both popular perception and a considerable body of older criminological writing, with opportunity for concealment. This intuition proved durable enough to shape municipal lighting policy for generations largely without rigorous empirical testing. When more careful, controlled research on the actual relationship between street-lighting levels and crime rates began to accumulate, particularly from the latter part of the twentieth century onward, the results proved considerably more mixed and complicated than the simple intuitive case would suggest: some studies have found modest reductions in certain categories of crime associated with improved street lighting, particularly for crimes that depend on concealment for opportunity, while others have found negligible or statistically insignificant effects, and a smaller number of studies have even suggested that excessive lighting, by improving visibility for potential offenders scouting a location or by creating harsh glare and shadow patterns that actually reduce a pedestrian's or security camera's effective visibility, can in some specific circumstances work against its intended purpose.

This is not a claim that lighting has no legitimate safety role; adequate illumination for pedestrian and vehicle safety, sufficient visibility for people to navigate their environment and recognize hazards, and reasonable lighting to support legitimate nighttime activity all remain genuinely well-supported goals. The point, relevant to this chapter's broader argument, is narrower: the specific, open-ended assumption that more brightness is always better for safety, used for over three centuries to justify essentially unlimited lighting expansion with little attention to actual necessity or diminishing returns, rests on considerably weaker empirical footing than its persistent use as a policy justification would suggest, and the gap between that persistent justification and the more equivocal underlying evidence is itself a recurring theme this book will return to in its final, solutions-oriented chapter.

Times Square as a case study in accumulated brightness

Before turning fully to Las Vegas, it is worth pausing briefly on Times Square in New York City, both because its own transformation illustrates the same accumulation dynamic in a different geographic and commercial context, and because it offers a useful contrast with the purpose-built, resort-driven brightness of Las Vegas discussed below.

Times Square's illuminated advertising signage dates to the earliest years of electrification, with large electric signs appearing along Broadway by the first decade of the twentieth century, and the area's association with spectacular nighttime illumination — eventually earning it the popular nickname "the Great White Way" — solidified over subsequent decades as competing advertisers, theater owners, and later corporations progressively added larger, brighter, and eventually animated and video-based signage in an accumulating competitive escalation with no obvious stopping point. Unlike Las Vegas, where illumination was built into a small number of large resort developments as a deliberate, unified marketing strategy from the outset, Times Square's brightness accumulated gradually, over more than a century, through the largely uncoordinated decisions of many separate advertisers and property owners, each adding their own increasingly elaborate signage in competition with neighbors rather than as part of any single coordinated plan.

The cumulative result, by the twenty-first century, is a comparably intense concentration of artificial light to anything found in Las Vegas, achieved through an entirely different institutional pathway: not a handful of casino operators competing directly with each other, but dozens of separate advertisers, over more than a hundred years, each adding incrementally to a shared visual environment none of them individually controlled or was responsible for as a whole. This distinction matters for the policy discussion in Chapter 15, because it illustrates that extreme concentrated light pollution can arise through genuinely different institutional mechanisms — centralized corporate decision-making in one case, decentralized, accumulated competitive escalation in the other — meaning that any policy response capable of addressing the full range of real-world light pollution sources needs to be flexible enough to address both patterns, not simply the more visually dramatic, single-actor case that a location like the Luxor beam represents.

The view from orbit

It is worth closing this chapter by noting a piece of evidence that did not exist for most of the period this chapter describes, but that has since become one of the most visually persuasive pieces of evidence for everything this chapter has argued: nighttime satellite and International Space Station photography of Earth, which has been collected systematically since the latter decades of the twentieth century and has become dramatically higher resolution and more widely available in the decades since. These images make directly, visually obvious what the preceding pages have described in narrative form: the world's major metropolitan regions stand out from orbit as distinct, brightly glowing clusters against a darker surrounding landscape, with certain very specific locations — the Las Vegas Strip prominent among them, given the concentrated intensity of lighting in a comparatively compact area surrounded by genuinely dark desert — appearing disproportionately bright relative to their physical size and population, a visual signature directly attributable to the specific historical and competitive dynamics this chapter has traced.

This orbital vantage point will recur, in a very different context, in the second half of this book, once the discussion turns from artificial light produced on the ground to artificial light produced by objects in orbit themselves — a shift in perspective, quite literally, from looking down at Earth's light pollution to looking up at a newer form of light pollution generated in orbit and directed back down toward Earth's surface. Chapter 5, immediately following, provides the tools necessary to describe and measure precisely how severe the ground-based version of that problem, described narratively throughout this chapter, has actually become.

Drive-in culture and the illuminated commercial strip

Beyond the specific examples of neon signage, floodlit skyscrapers, and highway lighting already discussed, the mid-twentieth-century rise of automobile-centered commercial development in the United States produced its own distinctive contribution to the era's lighting culture: the illuminated roadside commercial strip, populated by drive-in restaurants, motels, gas stations, and eventually shopping centers, each competing for the attention of drivers passing at speed, and each therefore designed, from the outset, around exactly the same attention-capturing logic already described for neon and floodlit architecture, simply applied at the smaller scale of an individual roadside business rather than an entire skyscraper or casino resort.

This roadside commercial lighting culture is worth noting specifically because of its sheer geographic extent: unlike the concentrated, spectacular lighting of a single downtown core or entertainment district, illuminated commercial strips extended, by the mid-to-late twentieth century, along thousands of miles of American highways and arterial roads, each individual business's signage modest by comparison to a Las Vegas casino, but the aggregate effect, spread across an entire continent's worth of roadside commercial development, contributing substantially to the overall diffusion of artificial lighting far beyond traditional city centers and into what had previously been comparatively dark suburban and exurban corridors, reinforcing and extending the highway-lighting-driven sky-brightening trend already discussed earlier in this chapter.

Parking lot lighting: an underappreciated contributor

A further, frequently underappreciated category of postwar lighting expansion deserving specific mention is parking lot and parking structure lighting, which grew enormously in both extent and intensity alongside the automobile-centered commercial and residential development patterns that came to dominate much of postwar American urban and suburban growth. Large retail centers, office parks, and multi-family residential developments each typically require substantial dedicated parking area, and the lighting standards applied to these areas, generally driven by liability and security considerations similar to those discussed earlier in this chapter's examination of the crime-and-lighting assumption, have historically tended toward high, uniform brightness levels maintained continuously throughout the entire night, regardless of actual usage levels during quieter overnight hours — precisely the pattern later addressed by the adaptive and motion-sensor lighting solutions discussed in Chapter 15.

The cumulative geographic extent of parking lot lighting specifically is considerable: a large suburban shopping mall or big-box retail development can involve parking areas covering many acres, each illuminated to a fairly high and uniform brightness standard essentially uninterrupted from dusk to dawn, and the aggregate area covered by this category of lighting across the full extent of postwar suburban commercial development in the United States alone represents a substantial fraction of the total artificial lighting footprint discussed throughout this chapter, despite receiving considerably less popular or scholarly attention than the more visually dramatic examples — neon signage, floodlit skyscrapers, the Luxor beam — this chapter has otherwise emphasized.

Illuminated signage regulation, early attempts

It is worth noting that a small number of American cities did attempt, even during the mid-twentieth-century heyday of illuminated commercial signage this chapter describes, to impose some regulatory limits on sign brightness and size, generally motivated by traffic-safety concerns (an excessively bright or distracting sign posing a hazard to passing drivers) or by aesthetic and historic-preservation concerns in specific designated districts, rather than by any concern for astronomical or ecological impact. These early regulatory efforts were typically narrow in scope, applied inconsistently, and rarely if ever framed in terms of sky brightness at all, illustrating that even where some lighting regulation existed during this period, it addressed a fundamentally different set of concerns than the dark-sky-specific ordinances that would not begin appearing until Flagstaff's pioneering 1958 ordinance, discussed at length in Chapter 15.

Film and popular culture's role in mythologizing the illuminated city

It is worth noting that twentieth-century film, photography, and popular culture played a significant role in cementing brightly lit cityscapes — and Las Vegas specifically — as symbols of glamour, modernity, and excitement, reinforcing rather than questioning the brightness-as-progress logic this chapter has traced. Countless films, postcards, and later, social media images have used the neon-lit strip or the floodlit skyline as visual shorthand for excitement and possibility, a cultural reinforcement mechanism that likely did as much as any direct economic incentive to normalize, and even glamorize, the specific kind of unrestrained nighttime brightness this chapter has described as a defining feature of the mid-to-late twentieth-century city.

Airport beacon lighting as a lesser-known category

Beyond the urban and highway lighting categories already discussed, the mid-twentieth-century expansion of commercial aviation introduced its own distinctive category of powerful, often continuously rotating beacon lighting at airports nationwide, justified purely by aviation-safety necessity but nonetheless contributing, at hundreds of individual airport sites, an additional, geographically dispersed source of concentrated nighttime brightness whose cumulative regional contribution to skyglow, while smaller than the urban and highway sources this chapter has emphasized, is rarely accounted for in popular discussions of the era's lighting expansion.

Municipal boosterism and world's fair afterglow

Cities that hosted the illuminated world's fairs discussed in Chapter 3 frequently sought to sustain some version of that fairground brilliance permanently afterward, funding ongoing illuminated civic monuments, bridges, and public buildings specifically to preserve the reputational glow, so to speak, of having once hosted a celebrated exposition, a civic-pride dynamic that further entrenched brightness as a marker of a city's ongoing modernity and status long after any specific fair had closed.

Chapter 5 — Measuring Darkness

For most of the history recounted in the preceding chapters, the disappearance of the stars over any given city occurred without anyone actually measuring it. There was no standard unit of sky brightness, no systematic monitoring network, no way to compare one location's night sky to another's except through subjective, unreliable personal impression. That began to change in the second half of the twentieth century, as amateur and professional astronomers, increasingly frustrated by the loss of observing conditions near major cities, developed the first serious tools for quantifying exactly how bright — or how dark — a given night sky actually is. This chapter introduces the most important and enduring of those tools: the Bortle Dark-Sky Scale.

Why "just drive out of the city" stopped working

Every experienced amateur astronomer of the mid-twentieth century understood, informally, that observing conditions improved with distance from a city, and the practical solution to deteriorating urban skies was simply to drive somewhere darker for serious observing. For decades this approach worked reasonably well, because a modest drive — twenty, thirty, fifty miles — was generally sufficient to escape a single city's direct glow and reach conditions close to a natural dark sky.

As the previous chapter described, however, postwar suburban sprawl and highway-linked commercial development steadily filled in what had once been genuinely dark countryside between cities, and the skyglow domes over individual metropolitan areas grew not only brighter but physically larger, extending farther beyond a city's official boundaries into what had previously been buffer zones of true darkness. By the latter decades of the twentieth century, observers in the densely populated eastern United States, most of western Europe, and comparable regions elsewhere increasingly found that no realistic drive was sufficient to fully escape the combined skyglow of multiple overlapping metropolitan areas — the light domes of separate cities had begun to merge into continuous regional skyglow, a phenomenon now clearly visible in satellite-derived light pollution maps covering the eastern seaboard of North America, most of western and central Europe, and Japan's Kanto plain, among other densely developed regions.

This changing reality created practical demand for a standardized way to describe and compare observing conditions, both so that astronomers could communicate precisely about a given site's quality, and so that the deteriorating trend could be documented with more rigor than anecdote.

John Bortle's scale

In 2001, amateur astronomer John E. Bortle published a nine-level scale in Sky & Telescope magazine, subsequently known as the Bortle Dark-Sky Scale, intended to give observers a standardized, descriptive framework for classifying the darkness of a given observing site based on what is actually visible to the naked eye and through a modest telescope, rather than a raw physical brightness measurement alone. The scale's enduring popularity, more than two decades after its publication, rests substantially on this descriptive, experiential approach: rather than requiring specialized equipment, an experienced observer can classify a site reasonably accurately simply by comparing what they see against Bortle's written descriptions of each class.

At the darkest end of the scale, Class 1 describes a truly pristine sky: the Milky Way, zodiacal light, and airglow are all readily visible and can even cast faint shadows on the ground; the zodiacal band extends across most of the sky; naked-eye limiting magnitude — the faintest star visible without optical aid — reaches roughly magnitude 7.6 to 8, meaning several thousand individual stars are visible simultaneously. Class 1 sites today are rare and typically found only in remote deserts, high-altitude locations far from any development, and a handful of protected areas specifically managed to preserve these conditions.

Class 2 and 3, still very dark, begin to show the first subtle effects of distant artificial light: a faint glow may be visible low on the horizon in the direction of the nearest significant town or city, though the zenith and most of the sky remain effectively pristine, and the Milky Way still displays rich structure and detail. Many established dark-sky preserves and remote rural observing sites fall into this range.

Class 4, described by Bortle as a "rural/suburban transition," marks the point at which light pollution becomes clearly noticeable in multiple directions rather than a single distant glow, with the Milky Way still obvious overhead but showing noticeably less structure and fewer of the finer details visible at darker sites. This is roughly the condition many rural and small-town observers experience today, a marked step down from Class 1 or 2 but still providing a genuinely useful and satisfying view of the night sky for both casual stargazing and serious amateur astronomy.

Class 5, “suburban sky,” is where the scale crosses into territory familiar to a very large fraction of the world’s population: skyglow is readily apparent in most or all directions, the Milky Way is at best faintly visible, appearing washed out and lacking the detailed structure so striking at darker sites, and the naked-eye limiting magnitude drops to around 5.2, cutting the number of simultaneously visible stars to a small fraction of the darkest-sky count.

Class 6 and 7 — “bright suburban” and “suburban/urban transition” — describe conditions in which the Milky Way is difficult or impossible to see at all without significant effort and dark adaptation, skyglow is bright enough to be the dominant visual feature of the sky in most directions, and only the brightest stars and a handful of prominent constellations remain easily recognizable to a casual observer. This range describes the observing conditions of most mid-sized cities and the immediately surrounding suburbs.

At the brightest end, Class 8 and 9 — “city sky” and “inner-city sky” — describe conditions in which the sky itself appears to glow a distinct color (often orange or yellowish-white, depending on the dominant type of lighting used locally), the Milky Way is entirely invisible, and only a limited selection of the brightest stars and planets remain visible at all, with naked-eye limiting magnitude often falling to 4 or below — meaning the vast majority of stars visible to the naked eye anywhere else on the scale simply cannot be seen from these locations, regardless of how clear the atmosphere is or how dark-adapted the observer's eyes become.

Watching the sky disappear, step by step

The value of the Bortle scale for this book's purposes lies precisely in this capacity for direct, step-by-step comparison: the same patch of sky, the same constellation, the same section of the Milky Way, looks measurably and describably different at each successive class, allowing a reader to trace, in concrete visual terms, exactly what is lost at each stage between a pristine Class 1 site and an inner-city Class 9 location. A useful mental exercise, and one this book will return to at intervals, is to imagine the same clear, moonless night, observed from progressively brighter locations along this ladder: a remote desert (Class 1 or 2), a small rural town (Class 4), a suburb of a major city (Class 5 or 6), and finally a dense urban core (Class 8 or 9). The stars have not moved and the atmosphere has not changed; only the artificial light scattered upward from the ground below has increased, and with each step, a further layer of the sky's structure and detail is erased from human perception, even though every star that was there at Class 1 is, physically, still there, still emitting the same light, simply overwhelmed by the brighter background against which it must now be distinguished.

Skyglow: the physical mechanism

Understanding why artificial ground lighting affects the sky at all — as opposed to simply illuminating the ground and stopping there — requires understanding the physical mechanism astronomers call skyglow. Artificial light directed upward, whether from unshielded streetlights, floodlit buildings, illuminated signage, or any other source, does not simply travel into space and disappear. It encounters the atmosphere: molecules of nitrogen and oxygen, water vapor, and, especially in and around urban areas, airborne particulates from combustion, dust, and pollution. These particles and molecules scatter the incoming artificial light in all directions, including back down toward the ground, in essentially the same physical process (Rayleigh and Mie scattering) that makes the daytime sky blue by scattering sunlight. The practical result is that a city's artificial light does not stay confined to the volume of air directly above it; it spreads, scattered and re-scattered, forming a dome of artificially brightened sky that can extend, for a sufficiently large city, many tens of kilometers beyond the city's actual boundaries — which is precisely why observers found that "driving out of town" became progressively less effective as cities and their light domes grew larger across the twentieth century.

The amount of skyglow a given light source produces depends heavily on factors within human control: the total light output, obviously, but also the proportion of that output directed above the horizontal (uplight, which contributes to skyglow with no corresponding benefit to ground-level visibility), the color temperature of the light source (bluer light scatters more effectively than warmer, redder light, a detail with significant implications for the LED lighting transition discussed in the final chapter of this book), and local atmospheric conditions such as humidity and particulate pollution, both of which increase scattering and therefore worsen skyglow for a given amount of artificial light. This list of controllable factors is worth noting here because it previews the central argument of the solutions chapter later in this book: skyglow, unlike some other forms of environmental harm, is substantially reducible through relatively modest engineering and policy changes, without necessarily requiring cities to simply turn off their lights.

Instruments beyond the eye

While the Bortle scale's descriptive, naked-eye approach remains extremely useful and widely referenced, it has been supplemented over the past two decades by increasingly precise instrumental measurement. Handheld sky quality meters, first becoming widely available to amateur astronomers and researchers in the mid-2000s, measure sky brightness directly in astronomically standard units and allow precise, repeatable, weather-independent comparison between sites and over time at a single site — critical for detecting gradual worsening trends that might be difficult for a human observer to notice reliably from memory alone. Satellite-based measurement, discussed in more detail in the next chapter, has more recently allowed researchers to map artificial light output across the entire surface of the Earth from orbit, producing the now-familiar global light pollution maps that make the scale and geographic extent of the problem immediately visible in a way no ground-based measurement network could easily replicate.

Together, the Bortle scale's descriptive framework and these more recent instrumental tools give this book's remaining chapters a vocabulary and a set of measurements precise enough to move beyond anecdote: to say, with some confidence, how much artificial light has increased over a given period, how far a given city's skyglow dome now extends, and how close to disappearing entirely the naked-eye Milky Way has become for the majority of the world's population — the specific subject of the next chapter.

Reading the scale in practice

To make the Bortle scale's nine categories more concrete, it helps to walk through how an experienced observer actually applies it in the field, since the scale was deliberately designed around observable phenomena rather than abstract numerical thresholds alone. An observer arriving at an unfamiliar site after dark, once fully dark-adapted (a process that itself takes twenty to thirty minutes of avoiding bright light, since human night vision continues improving well beyond the first few minutes many people assume is sufficient), typically begins by assessing the Milky Way's visibility and structure overhead, then checks whether the zodiacal light is detectable near the horizon before dawn or after dusk, then estimates naked-eye limiting magnitude by identifying the faintest stars visible in a specific, well-known reference area of sky (several standard star fields are commonly used for exactly this purpose, since they contain stars of precisely known brightness spanning the relevant range), and finally notes the presence, direction, and intensity of any visible glow from distant artificial sources along the horizon.

This combination of observations, taken together, typically allows an experienced observer to place a site confidently within one or two adjacent Bortle classes, and cross-referencing against instrumental sky-quality-meter readings (discussed further below) generally confirms this naked-eye classification with good consistency, which is part of why the scale has remained the standard descriptive reference across the amateur and professional astronomical community for more than two decades, despite the availability of more precise instrumental alternatives for over half of that period.

A worked example: the same sky, four locations

Consider a specific, concrete illustration that captures the core argument of this chapter and this book as directly as possible: the constellation Sagittarius, which sits along one of the richest and most structurally detailed sections of the Milky Way as seen from Earth, containing the direction toward our galaxy's crowded central bulge.

From a Class 1 or 2 site — a remote desert or high-altitude location with minimal nearby development — an observer looking toward Sagittarius on a clear summer night sees not merely a scattering of individual stars but an obviously three-dimensional, richly textured structure: bright star clouds standing out clearly from darker, dust-obscured lanes running between them, the overall band of the Milky Way clearly brighter and more structured in this direction than elsewhere in the sky, and, with sufficiently dark adaptation, the naked eye can begin to resolve individual bright clusters and nebulae within this region as faint, fuzzy patches distinct from ordinary point-like stars.

From a Class 4 or 5 site — a small town or a suburb of a mid-sized city — the same constellation, on the same night, at the same time, shows a recognizably fainter and less detailed version of the same view: the Milky Way's band remains visible as a diffuse glow in the right conditions, but the fine dust-lane structure and the distinction between brighter star clouds and darker intervening regions has become considerably harder to discern, and the fainter clusters and nebulae that were faintly visible from the darker site have generally dropped below what the naked eye can detect against the brighter sky background.

From a Class 7 or 8 site — a larger city or its immediate surroundings — the Milky Way in this same direction has typically become undetectable to the naked eye entirely, and only the very brightest individual stars within the constellation remain visible at all, appearing as a handful of isolated points against an otherwise essentially blank, artificially brightened sky, with none of the structural richness described in the darker examples above remaining perceptible in any way.

This is not a hypothetical illustration; it describes, accurately, an experience that anyone with access to both a genuinely dark site and a well-lit city, and the patience to observe the same patch of sky from both locations, can personally verify. It is precisely this kind of direct, repeatable, personally verifiable comparison that has made the Bortle scale as durable and as rhetorically persuasive a tool as it has become, both within the astronomical community and, increasingly, in public communication about light pollution aimed at a broader, non-specialist audience.

Sky quality meters and the move to instrumental precision

While the descriptive, naked-eye Bortle scale remains widely used precisely because it requires no specialized equipment, the development of affordable, portable sky quality meters beginning in the mid-2000s added a valuable complementary tool: a handheld instrument, typically the size of a small flashlight, that measures the brightness of the night sky directly overhead in standardized units (typically expressed as magnitudes per square arcsecond, a logarithmic astronomical brightness unit) and produces a precise, repeatable numerical reading independent of any individual observer's visual acuity, dark adaptation, or subjective judgment.

This instrumental precision matters for two related reasons that go beyond simple convenience. First, it allows meaningful comparison and trend-tracking over time at a single site with a rigor that purely visual, memory-based comparison cannot match; a sky quality meter reading taken tonight can be compared directly and quantitatively to a reading taken at the same location a decade earlier, in a way that "it seemed darker when I was a kid" cannot support with equivalent scientific rigor. Second, it enables exactly the kind of large-scale, standardized data collection that underlies the citizen-science monitoring programs discussed in the next chapter, since a network of observers using consistent, calibrated instruments across many different locations can build a dataset suitable for rigorous statistical analysis of regional and global light pollution trends in a way that purely qualitative, Bortle-scale-only observations, however useful for individual site description, cannot as readily support at that larger analytical scale.

Together, the descriptive Bortle framework and the quantitative sky-quality-meter data it complements give the next chapter's discussion of global light pollution trends a solid empirical foundation: not merely impressionistic accounts of skies getting worse, but a rigorous, decades-deep body of standardized measurement capable of showing, with real statistical confidence, exactly how much worse, how quickly, and where.

The Bortle scale's limitations, honestly stated

No discussion of the Bortle scale would be complete without acknowledging its genuine limitations, which the astronomical community itself has openly discussed in the decades since its publication. The scale's descriptive categories, while useful and broadly consistent in application among experienced observers, inevitably involve some degree of subjective judgment, particularly at the boundaries between adjacent classes, and different experienced observers applying the scale to the same site on the same night have occasionally disagreed by a class or so, a limitation that instrumental sky-quality-meter readings, discussed later in this chapter, help but do not entirely eliminate, since instrumental readings measure a somewhat different physical quantity (overall sky radiance) than the specific perceptual criteria (Milky Way structure, limiting magnitude, and similar naked-eye phenomena) the Bortle scale's written descriptions actually specify.

Additionally, the scale was originally developed with mid-latitude Northern Hemisphere observing conditions primarily in mind, and its specific descriptive benchmarks — which constellations are used as reference points for limiting-magnitude estimation, for instance — translate somewhat imperfectly to Southern Hemisphere or extreme-latitude observing conditions, a limitation subsequent astronomical writing has generally addressed through supplementary reference material rather than a formal revision of the scale itself. None of these limitations undermine the scale's core value as a broadly consistent, widely understood, easily communicated descriptive framework, and it remains, more than two decades after publication, the standard reference used throughout the amateur and professional astronomical community, as well as in the dark-sky advocacy and policy contexts discussed in Chapter 9 — but the limitations are worth naming plainly, in keeping with this book's broader commitment to representing the underlying science with appropriate precision rather than overstating the certainty of any single measurement tool.

Skyglow's directional character

One further technical point about skyglow deserves elaboration here: sky brightening from a given light source is not uniform across the sky, but concentrated in the direction of, and generally decreasing with angular distance from, the source itself. An observer positioned with a large city to their north, for instance, will typically observe the most severe skyglow low on the northern horizon, with progressively darker conditions moving away from that direction toward the zenith and the southern sky, though the zenith itself is rarely entirely unaffected in any location within a reasonable distance of significant development, since scattered light disperses considerably more broadly than a simple direct-line-of-sight model would suggest.

This directional character has a practical implication directly relevant to the observing-site selection practices discussed throughout the dark-sky preservation literature: an observing site's overall Bortle classification is generally driven by its worst, most light-polluted direction, but a site with excellent darkness in most directions and only a single significant nearby light source in one specific direction can still support high-quality observation of objects located away from that specific direction, a nuance that experienced observers and dark-sky-site evaluators take into account when assessing a candidate location's suitability for specific observing projects, rather than relying on a single overall classification alone.

Photography and the popular discovery of skyglow

Long-exposure nighttime photography, becoming increasingly accessible to amateur photographers from the mid-twentieth century onward, played an underappreciated role in making skyglow visible and legible to a broader public audience well before formal scientific measurement tools were widely available. A photographic exposure of several seconds or more, capturing far more light than the human eye can integrate in real time, reveals the characteristic orange-white glow of a distant city's skyglow far more vividly and unambiguously than casual naked-eye observation typically does, and photographs of this kind, increasingly shared and compared among amateur astronomers and, later, on early internet forums dedicated to astrophotography, helped build informal public awareness of the skyglow phenomenon well in advance of the more formal Bortle scale and satellite-mapping tools this chapter has otherwise emphasized.

The scale as public communication tool

Beyond its technical use among astronomers, the Bortle scale has increasingly become a public-communication device in its own right, used by park interpretive programs, dark-sky tourism marketing, and popular science journalism to give a general audience some quantitative purchase on what would otherwise be a purely qualitative claim about "how dark" a given sky is. This broader public adoption, well beyond the scale's original amateur-astronomy audience, is itself a sign of how thoroughly the underlying concept of measurable, comparable sky darkness has entered general public discourse over the two decades since the scale's original publication, a shift in public literacy this book's own use of the scale, throughout its historical and analytical chapters alike, has depended on and hopes to further extend.

Color-sensitive brightness perception

It bears repeating, with slightly more precision than earlier in this chapter, that human perception of sky brightness is not a simple linear function of total photon count; the eye's differing sensitivity across the visible spectrum, and its shift toward blue-sensitive rod vision under low-light, dark-adapted conditions, means two skies with identical instrumentally measured radiance but different dominant color temperatures can appear meaningfully different in perceived brightness and quality to a human observer, a subtlety directly relevant to the LED color-temperature discussion this book returns to in its final chapter.

Training new observers

Astronomical societies and dark-sky organizations increasingly run dedicated training sessions teaching newcomers how to apply the Bortle scale and use sky quality meters correctly, recognizing that consistent, credible citizen data collection depends on reasonably well-trained observers following standardized procedures, an investment in observer training that has measurably improved the quality and consistency of the citizen-contributed data underlying much of the trend research discussed in the following chapter.

Chapter 6 — Where the Stars Went

Equipped, from the previous chapter, with a vocabulary for describing sky darkness precisely, we can now ask a more specific and more unsettling question: how much of the world's population has actually lost meaningful access to a dark night sky, how fast did that loss occur, and what does the pattern look like when mapped across the globe rather than described city by city?

The global light pollution atlas

The most influential single piece of research addressing this question directly is the "World Atlas of Artificial Night Sky Brightness," first published in 2001 and substantially updated in 2016 by an international team of researchers using high-resolution satellite imagery of nighttime artificial light output combined with atmospheric modeling to estimate actual sky brightness, as it would be experienced by an observer on the ground, across virtually the entire populated surface of the Earth. The 2016 update, published in the journal Science Advances and widely covered in general media at the time, produced a specific and often-quoted set of figures that crystallized, for a broad public audience, a problem astronomers and dark-sky advocates had been describing qualitatively for decades.

According to that study, roughly 80 percent of the world's population, and more than 99 percent of the population of the United States and Europe specifically, lives under sky conditions affected by measurable light pollution. More strikingly, the same research estimated that approximately one-third of humanity as a whole — and roughly 80 percent of North Americans specifically — can no longer see the Milky Way from their home location at all, the artificial skyglow from nearby development having brightened the local sky past the threshold at which the galaxy's diffuse structure remains perceptible to the naked eye. For a very large fraction of the world's most economically developed regions, in other words, the sky described in Chapter 1 — the Milky Way as visible structure, part of ordinary nighttime experience — is not merely diminished. It is gone entirely, not as an occasional inconvenience on a hazy night, but as a permanent, year-round condition of where these populations live.

Regional variation and the merging of light domes

The global picture masks enormous regional variation, and examining that variation helps explain both how the problem developed historically and where meaningful dark sky still survives. Densely populated, heavily industrialized, and heavily lit regions — the eastern United States from roughly Washington, D.C. through Boston, most of western and central Europe from the British Isles through Germany and into northern Italy, Japan's Kanto and Kansai plains, and rapidly urbanizing regions of China, India, and Southeast Asia — show not merely bright localized skyglow around individual cities but continuous or near-continuous regional light domes in which the artificial brightening from adjacent metropolitan areas has effectively merged, eliminating the dark buffer zones that separated those cities' skyglow as recently as the mid-twentieth century.

By contrast, large areas of the American West, significant portions of the Australian interior, the Sahara and other major deserts, much of Siberia, parts of the Amazon basin, and a scattering of other sparsely populated regions retain skies close to or at the pristine, Class 1 or 2 end of the Bortle scale — not because of any specific conservation effort in most cases, simply as a byproduct of low population density and correspondingly limited artificial lighting infrastructure. These remaining dark regions are not evenly distributed relative to where people actually live, however; the same 2016 atlas found that for a resident of, say, Belgium or the densely populated northeastern United States, reaching genuinely dark, Class 1-equivalent skies requires travel of many hundreds of kilometers in almost any direction — a dramatically different situation from a resident of Wyoming or the Australian outback, for whom dark skies may be available within a short local drive.

How fast has this actually happened?

The historical chapters of this book have already traced the century-and-a-half arc from gaslight through neon and floodlighting to the contemporary LED-lit metropolis, but it is worth being precise here about the pace of change specifically as it is reflected in measured sky brightness, because the recent rate of change turns out to be, if anything, more alarming than the long historical arc alone would suggest.

Research published in 2023 in the journal Science, analyzing over a decade of citizen-scientist observations of naked-eye star visibility collected through the long-running Globe at Night program, found that sky brightness as experienced by ground-based observers had increased at an average rate of approximately 10 percent per year between 2011 and 2022 — a rate substantially faster than satellite measurements of total artificial light output over the same period would have predicted, a discrepancy the researchers attributed at least in part to the ongoing global transition from older sodium-vapor lighting to LED fixtures, discussed further in Chapter 15, whose different spectral characteristics appear to produce more sky brightening, as perceived by the human eye and standard sky-brightness instruments, per unit of measured light output than the older technology it is replacing. A sustained 10 percent annual increase compounds extremely quickly: at that rate, a location's sky brightness would roughly double in under eight years, meaning a child born into a given level of light pollution could plausibly experience a dramatically darker starting condition for their childhood memories of the night sky than their own children would experience at the same age only a decade or two later, within a single extended family's living memory.

This finding — that the rate of change in recent years may be accelerating rather than merely continuing a steady historical trend — is significant for the argument of this book because it complicates any comfortable assumption that the light pollution problem, having already caused most of its damage across the twentieth century, might now be leveling off as cities reach some kind of lighting saturation point. The evidence instead suggests the opposite: continued global urbanization, the ongoing global LED conversion, and increasing use of decorative, security, and advertising lighting in rapidly developing regions of the world are combining to keep sky brightness increasing at a pace that, if sustained, would leave even fewer dark-sky refuges accessible within a human lifetime than currently exist.

What a "never seen the Milky Way" childhood actually means

It is worth pausing to consider, concretely, what it means that a large and growing fraction of children in economically developed countries are being raised with no expectation of ever seeing the Milky Way as part of ordinary experience — not because they have never had the opportunity to travel to a dark location, necessarily, but because the default condition of the sky above their home, school, and daily environment offers no visible hint that the galaxy has a visible structure at all. A child growing up in most major metropolitan areas today can reasonably reach adulthood, complete a full education including whatever astronomy content their science curriculum includes, and never once have the direct, unmediated experience their great-grandparents would have had simply by stepping outside on an ordinary clear night.

This is a different kind of loss than the more commonly discussed environmental and ecological consequences addressed in the coming chapters, though it is connected to them. It is a loss of a specific kind of cultural inheritance — the night sky as shared, ordinary, universally accessible experience, described in Chapter 1 as functioning almost like public infrastructure for the whole of human history until shockingly recently — and its disappearance from daily life, even as the technical means to travel and witness dark skies deliberately remains available to those with the resources, time, and inclination to do so, represents a significant narrowing of what counts as normal, expected, ambient human experience. Chapter 8 will examine the psychological and cultural dimensions of this loss in more detail; this chapter's purpose has been simply to establish, with as much precision as the available research allows, the scale and pace of the underlying phenomenon: not a modest dimming at the margins, but the substantial or complete disappearance of the visible Milky Way for roughly a third of the human species, achieved within considerably less than a single century, and continuing to worsen at a pace that shows no clear sign of slowing.

The next chapter turns to a different but closely related set of consequences: what this same transformation of the night has done not to human cultural experience, but to the animal species that share the planet with us, many of which depend on natural patterns of light and dark in ways far more direct, and in some cases far more immediately consequential, than anything discussed so far.

Globe at Night: the citizen-science backbone

Much of the trend data referenced in this chapter, particularly the finding that ground-observed sky brightness has been increasing considerably faster than satellite-measured light output alone would predict, comes from a specific, long-running citizen-science program worth describing in more detail: Globe at Night, coordinated since the mid-2000s by the National Optical Astronomy Observatory and international partners, which asks members of the public worldwide to make a simple naked-eye observation — comparing the visibility of a specific constellation against a set of reference star charts showing how that constellation would appear under varying levels of light pollution — and submit that observation, along with their location, through a standardized online reporting system.

The program's design deliberately trades the precision of a single expert instrumental reading for the sheer geographic scale and volume achievable by inviting broad public participation: hundreds of thousands of individual observations have been submitted from locations across every populated continent since the program's inception, creating a dataset with genuinely global coverage that would be prohibitively expensive to achieve through professional instrumental monitoring alone, even though any individual citizen-scientist observation is necessarily less precise than a calibrated sky-quality-meter reading taken by a trained observer. The 2023 Science study referenced earlier in this chapter, finding a roughly 10 percent annual increase in perceived sky brightness between 2011 and 2022, drew directly on this accumulated body of citizen-submitted observations, and its findings have been influential precisely because the underlying dataset's scale allowed researchers to detect a clear, statistically robust trend across a very large and geographically diverse sample, in a way that a smaller number of professional monitoring stations, however precise individually, might not have been able to establish with equivalent confidence.

Why satellite and ground measurements diverge

The apparent discrepancy this chapter has already noted — ground-observed brightness increasing faster than satellite-measured total light output — deserves a slightly fuller technical explanation, because it is not simply a data-quality artifact but reflects a genuine, physically meaningful limitation in how satellites measure artificial light from orbit.

The satellite imagery most commonly used to construct global light pollution maps, including the influential 2016 World Atlas referenced earlier in this chapter, relies on sensors calibrated to detect a specific range of light wavelengths, and critically, the sensors used for the majority of the historical satellite record have had limited or no sensitivity to the blue end of the visible spectrum — precisely the wavelength range that has become disproportionately more common as cities have replaced older sodium-vapor lighting with blue-enriched LED fixtures, a transition discussed at greater length in Chapter 15. This means that a city's LED conversion can measurably increase the actual sky brightness experienced by a ground-based human observer, who perceives the full visible spectrum including blue wavelengths, without producing an equivalent increase in the total light output as measured by satellite sensors that are comparatively blind to exactly the wavelength range where the increase is concentrated. Researchers studying this discrepancy have described it as a significant complicating factor for anyone relying solely on satellite data to track light pollution trends, since satellite data, however valuable for large-scale spatial mapping, may be systematically underestimating the true trajectory of human-experienced sky brightness precisely during the period of most rapid LED adoption, making ground-based citizen-science and instrumental monitoring, of the kind described above, an essential complement rather than a redundant alternative to satellite-based measurement.

Regional case studies: three cities, three trajectories

To make the abstract, global-scale statistics discussed earlier in this chapter more concrete, it is worth briefly considering three specific, well-documented regional cases that illustrate different trajectories within the overall global trend.

Seoul, South Korea, one of the most extensively documented cases of rapid, large-scale artificial-light growth tied to compressed economic development, saw its metropolitan light footprint expand dramatically across the closing decades of the twentieth century and the opening decades of the twenty-first, tracking the country's broader rapid industrialization and urbanization; satellite-derived brightness data for the Seoul metropolitan region shows this growth as one of the more dramatic examples of national-scale light pollution increase documented anywhere in available satellite records, illustrating how quickly a region's night sky can transform when rapid economic development and correspondingly rapid electrification and urban lighting expansion occur within a single generation, rather than the more gradual, multi-century pattern traced in earlier chapters for cities in longer-industrialized regions.

By contrast, certain long-industrialized regions of Western Europe, particularly parts of Germany and the United Kingdom that adopted relatively early and, in some cases, comparatively conservative outdoor lighting standards, show more modest recent growth rates in some satellite-tracked datasets than comparably sized rapidly developing regions elsewhere, though even these more slowly growing regions remain, in absolute terms, among the most light-polluted on the planet given more than a century of accumulated prior development, illustrating that "slower recent growth" and "genuinely dark sky" are two entirely different things, easily conflated if the underlying absolute baseline is not kept clearly in view.

And a small number of specific municipalities – Flagstaff, Arizona, discussed at length in Chapter 15, being the most thoroughly documented example – show measurably slower sky-brightness growth than population and commercial growth alone would predict, directly attributable to sustained, deliberate lighting-ordinance policy adopted and maintained over multiple decades, offering an important counterpoint to the otherwise fairly uniform picture of accelerating light pollution presented elsewhere in this chapter: the trend, however dominant, is not literally universal or inevitable everywhere, and specific, documented policy interventions have measurably altered it at the local level in at least some well-studied cases.

These three cases, taken together, previously seed a theme this book will return to directly in its final chapters: the global trend toward increasing sky brightness, while real, severe, and in recent years apparently accelerating, is not a uniform law of nature operating identically and irresistibly everywhere; it is the aggregate outcome of countless individual local decisions about lighting technology, urban development, and policy, and at least some of those decisions, made differently, have produced measurably different local outcomes, a fact whose full policy significance the book will develop later, once the orbital half of the light-pollution story has also been told.

Light pollution and social equity

An important dimension of the global distribution pattern discussed in this chapter, not yet directly addressed, concerns the relationship between light pollution exposure and socioeconomic factors within individual metropolitan regions. A number of studies examining the distribution of both terrestrial light pollution and associated health and environmental impacts within specific cities have found that lower-income and historically marginalized neighborhoods frequently experience measurably higher levels of certain forms of light pollution — from industrial and commercial lighting, elevated highway and arterial road density, and, in some documented cases, differential municipal investment in the kind of shielded, dark-sky-friendly fixture upgrades discussed in Chapter 15 — than wealthier neighborhoods within the same metropolitan region, a pattern broadly consistent with well-documented findings regarding other forms of environmental burden and their uneven distribution across socioeconomic lines within many cities.

This equity dimension matters for a full accounting of light pollution's human costs, discussed at greater length in the next chapter, because it means that the sleep-disruption and circadian-health concerns addressed there are not evenly distributed across a given population, but may disproportionately affect populations already facing other, compounding environmental and health burdens, a pattern that a number of public-health researchers examining environmental justice issues more broadly have argued deserves greater explicit attention within light-pollution research and policy specifically, rather than being treated as a purely aesthetic or astronomical concern with no meaningful equity dimension of its own.

Similarly, access to the dark-sky recreational and cultural experiences discussed in Chapter 9 — travel to certified dark-sky parks and reserves, participation in astronomy tourism — is itself unevenly distributed by income and geographic circumstance, meaning that the cultural and psychological costs of losing routine access to a visible night sky, discussed at length in the next chapter, may fall disproportionately on populations with the least practical ability to travel and recover that access through the kind of deliberate dark-sky tourism increasingly available to more affluent populations, a dimension of the overall light-pollution story that deserves to be named clearly even though, as with several other topics touched on in this chapter, the specific research base quantifying its full extent remains considerably less developed than the broader global distribution data this chapter has otherwise relied on.

Long-term projections, cautiously stated

Building on the accelerating recent trend data discussed earlier in this chapter, it is worth briefly noting what a small number of researchers have attempted in terms of longer-term projection, while maintaining appropriate caution about the inherent uncertainty involved in any such exercise. If the roughly 10 percent annual sky-brightness increase documented for the 2011-2022 period referenced earlier in this chapter were to continue unchanged for an additional decade or two — an assumption that should not be treated as a confident prediction, given the many factors, including the kind of policy interventions discussed in Chapter 15, that could plausibly alter this trajectory in either direction — the practical implication would be a further, very substantial reduction in the fraction of the world's population retaining any meaningful naked-eye access to the Milky Way beyond the already-diminished roughly two-thirds share estimated by the 2016 World Atlas discussed earlier in this chapter.

This kind of straightforward trend extrapolation is offered here not as a confident forecast, but as a further illustration of the same basic point this chapter has emphasized throughout: the currently available data describes an already severe and, on the most recent evidence, still-worsening situation, and the case for the kind of deliberate policy intervention discussed in this book's final chapters rests substantially on the observation that an already serious trend shows no clear sign, absent such intervention, of leveling off or reversing on its own.

Population growth versus per-capita lighting growth

It is worth distinguishing, in interpreting the global trend data discussed throughout this chapter, between sky-brightness growth attributable simply to more people existing and building more infrastructure, and growth attributable to increased per-capita lighting intensity independent of population change. Both factors are clearly contributing to the trends described in this chapter, but they carry somewhat different policy implications: population-driven growth is comparatively difficult to address through lighting policy alone, since it reflects broader demographic and development trends outside the scope of any lighting ordinance, whereas per-capita intensity growth —

brighter fixtures, more decorative and security lighting per household, more illuminated advertising per commercial establishment — is considerably more directly addressable through exactly the kind of fixture and ordinance-level interventions discussed in Chapter 15, since it reflects discretionary choices about how much light a given unit of population or development chooses to use, rather than the sheer fact of that population or development existing at all.

National mapping initiatives beyond the global atlas

Beyond the global World Atlas discussed at length in this chapter, a number of individual countries have developed their own more granular, nationally focused light pollution mapping initiatives, often achieving considerably higher spatial resolution than a global-scale dataset can practically support, and allowing more precise, locally actionable identification of specific light pollution sources and trends within a given country's borders. These national mapping efforts have, in a number of cases, directly informed the kind of municipal and national lighting policy discussed in Chapters 9 and 15, providing the specific, locally credible data policymakers in a given jurisdiction need to justify and design targeted interventions, complementing rather than replacing the broader, comparative value of the global-scale data this chapter has primarily relied on.

Cloud cover as a confounding variable

Researchers analyzing long-term sky-brightness trend data must also account for changing regional cloud-cover patterns, since clouds dramatically amplify skyglow by reflecting artificial ground light back downward rather than allowing it to disperse into the upper atmosphere and beyond, meaning some of the year-to-year variability in a given region's measured sky brightness reflects shifting weather patterns rather than genuine underlying change in artificial light output, a statistical complication the researchers behind the major studies this chapter has cited have generally addressed through multi-year averaging and careful weather-condition filtering.

Migration patterns and shifting monitoring baselines

Researchers tracking long-term sky-brightness change must also account for population migration and shifting regional development patterns, since a monitoring station's surrounding land use, and therefore its local light pollution baseline, can itself change significantly over a multi-decade monitoring period independent of any broader lighting-technology trend, a methodological complication the major longitudinal studies referenced throughout this chapter have generally addressed through careful site selection and, where necessary, exclusion of stations whose surrounding land use changed too substantially over the study period to support clean trend analysis.

Chapter 7 — The Animals Don't Know It's Artificial

Every ecological relationship discussed in this chapter rests on a single, simple fact: the daily and seasonal cycle of light and dark is one of the oldest and most reliable environmental signals on Earth, older than complex multicellular life itself, and an enormous number of species across every branch of the animal kingdom have built their behavior, physiology, and life cycles around it over hundreds of millions of years of evolution. Artificial light at night — the same phenomenon this book has so far discussed largely in terms of human experience and astronomy — does not read as a curiosity or an aesthetic loss to these species. It reads as a signal, and for many animals, an actively false and disorienting one, arriving on a timescale (roughly a century and a half of significant artificial lighting, and a much more compressed few decades of truly severe artificial lighting) far too short for evolutionary adaptation to catch up.

Sea turtle hatchlings: the clearest case study

Among the best-documented and most visually striking examples of artificial light disrupting animal behavior is the disorientation of sea turtle hatchlings on nesting beaches. Sea turtles of multiple species nest on beaches worldwide, laying eggs that incubate in beach sand for roughly two months before hatching, typically at night. For the entire evolutionary history of sea turtles, hatchlings emerging from a nest at night have relied on a simple and, until recently, entirely reliable cue to find the ocean: the natural night sky and horizon over open water is measurably brighter than the darker silhouette of dune vegetation and land behind the beach, and hatchlings instinctively crawl toward the brighter direction.

Artificial lighting from beachfront development — hotels, homes, streetlights, parking lots — inverts this cue completely. To a hatchling emerging onto a developed beach, artificial lights from buildings and roads behind the beach are frequently brighter than the natural sky-glow reflected off the ocean, and hatchlings, following the same instinct that has reliably guided their ancestors to the sea for millions of years, instead crawl inland, toward the artificial lights, away from the water. The consequences are severe and well documented by conservation researchers across nesting beaches worldwide: disoriented hatchlings exhaust themselves crawling in the wrong direction, become stranded in vegetation or on roads, fall prey to predators that would not otherwise encounter them in such numbers, or are killed directly by vehicle traffic, all before ever reaching the ocean. Conservation programs in Florida, a major nesting ground for several turtle species, have implemented specific lighting ordinances in coastal counties — requiring shielded, turtle-safe fixtures and seasonal lighting restrictions during nesting season — specifically to address this single, well-characterized mechanism, and monitoring data from beaches that have adopted such ordinances generally show measurably improved hatchling survival compared to unlit or poorly regulated comparison sites.

Migratory birds: collision and disorientation

A second major category of documented harm involves migratory birds, many species of which migrate at night, using a combination of celestial cues (including, for some species, orientation relative to the stars themselves, learned during a critical developmental period), the Earth's magnetic field, and geographic landmarks to navigate over enormous distances between breeding and wintering grounds. Artificial light — particularly the concentrated, upward-directed light of illuminated buildings, communication towers, and other tall urban structures — has been shown, through both direct observation and, increasingly, weather radar data repurposed by ornithologists to track bird movement, to attract and disorient nocturnally migrating birds passing near or over lit areas.

The mechanism appears to involve birds becoming drawn toward bright light sources, circling them repeatedly rather than continuing their migratory path, sometimes for hours, in a pattern that increases exhaustion, and in the specific case of illuminated skyscrapers and communication towers, results in substantial numbers of collision deaths as disoriented birds fly directly into glass facades or structural elements, particularly on nights with low cloud ceiling that concentrates migrating birds at lower altitudes closer to city light sources. Studies of building-collision mortality in major North American cities situated along significant migratory flyways — Chicago and New York being particularly well studied — have documented collision events involving many hundreds of birds at a single building on a single night during peak migration periods, and citywide "Lights Out" programs, which ask building managers to reduce or extinguish nonessential exterior and interior lighting during peak spring and fall migration windows, have become one of the more widely adopted and measurably effective light-pollution mitigation efforts specifically targeted at this problem, with radar and ground-survey data from participating cities showing meaningful reductions in both bird-attraction intensity and collision mortality on nights when lighting reduction commitments are actually honored.

Insects: the vacuum effect and a case study in miniature

Insects present perhaps the most visually familiar example of artificial light's behavioral effects, since the tendency of many flying insect species to gather around bright lights at night is a phenomenon most people have observed directly, around a porch light or streetlamp, without necessarily understanding its broader ecological significance. The precise mechanism behind this attraction remains an active area of scientific study — competing explanations include disruption of the insects' normal method of orienting flight relative to a distant, fixed light source such as the moon (which artificial lights, being much closer, distort into a spiraling flight path around the light itself), photoreceptor overstimulation, and simple failure of evolved escape or avoidance responses in the presence of an evolutionarily novel stimulus — but the practical ecological consequence is increasingly well documented regardless of mechanism: artificial light functions, in the language now commonly used by entomologists studying the phenomenon, as a population-scale "vacuum effect," drawing insects out of surrounding habitat and either killing them directly (through exhaustion, increased predation risk while circling a light source, or direct contact with hot fixtures) or disrupting their normal reproductive and foraging behavior badly enough to reduce local population viability over time.

This matters far beyond simple aesthetic annoyance because insects occupy foundational positions in nearly every terrestrial food web and pollination system on Earth, and a growing body of research — including long-term monitoring studies in parts of Europe documenting substantial declines in overall flying-insect biomass over recent decades — has identified artificial light at night as one of several likely contributing factors in broader insect population declines, alongside habitat loss, pesticide use, and climate change, though disentangling light pollution's specific independent contribution from these other simultaneous pressures remains genuinely difficult and is an active area of ongoing research.

The Luxor Sky Beam, introduced in Chapter 4, offers an unusually clean, almost laboratory-like illustration of this dynamic operating in miniature, precisely because it is a single, extremely powerful, well-documented light source situated in an otherwise dark desert environment with relatively few confounding light sources nearby. Observers and researchers examining the beam have documented enormous concentrations of insects drawn upward into the vertical column of light, sometimes described as visible from a considerable distance as a moving, swarming mass within the beam itself on warm nights. This concentration of insect prey, in turn, reliably attracts bats — several species of which have been observed foraging within and around the beam specifically to exploit this artificially concentrated food source — creating a small but genuine artificial ecosystem: a food web that exists, in this specific form and at this specific location, purely because of a light fixture built for architectural spectacle, with no other ecological function whatsoever. It is worth returning to this example here, rather than treating it merely as an entertaining anecdote from Chapter 4, because it demonstrates concretely and at a single, observable site something that is otherwise happening diffusely, and much less visibly, around billions of individual light fixtures worldwide: artificial light does not merely fail to affect wildlife: it actively restructures which species are present, in what numbers, and behaving in what ways, in the immediate vicinity of that light, with consequences that ripple outward through the surrounding food web in ways still only partially understood.

Broader effects across additional species and systems

Beyond these three especially well-documented cases, researchers have identified light pollution effects across a considerably wider range of species and ecological processes. Coral species in several reef systems have been shown to depend on natural moonlight cues to synchronize the precisely timed, mass spawning events characteristic of many reef-building corals, and artificial light pollution near coastal reef systems has been implicated in disrupting the timing precision of these events, with potential downstream consequences for coral reproduction success, though this remains an active area of research with fewer long-term studies than the sea turtle and bird literature. Amphibians, many of which rely on nighttime darkness for foraging, breeding call behavior, and predator avoidance, show measurable behavioral disruption in artificially lit environments in multiple documented field studies. Nocturnal mammals more broadly, from bats (beyond the specific opportunistic feeding behavior described above) to rodents to larger predators, frequently show altered activity patterns, habitat avoidance, or foraging behavior changes in response to artificial lighting, with effects that vary considerably by species — some species show increased vulnerability to predation under artificial light, while a smaller number of predator species appear to exploit artificially lit areas specifically because prey species are more visible or more disoriented there, illustrating that light pollution's ecological effects are not uniform but instead reshape the relative competitive and survival balance among different species sharing an ecosystem, typically in ways that were never intended or anticipated by whoever installed the light source responsible.

Even diurnal (day-active) species, intuitively less obviously affected by nighttime lighting, show measurable physiological and behavioral responses to artificial light at night in a number of documented studies, including disrupted sleep patterns, altered activity timing, and, in some bird species, effects on breeding timing and hormonal cycles tied to disrupted perception of day length — a reminder that the light-dark cycle functions as a signal relevant to essentially the entire animal kingdom, not merely to species that are active exclusively after dark.

An accidental, unregulated experiment

What unites all of the examples in this chapter is a common structural feature worth naming explicitly: none of this ecological disruption was intended, anticipated, or in most cases even studied at the time the relevant lighting infrastructure was installed.

Beachfront hotel lighting was installed for guest convenience and safety; skyscraper and communication-tower lighting was installed for aviation safety, security, or aesthetic effect; streetlights and highway lighting were installed, as earlier chapters described, on largely unexamined assumptions about crime prevention and driver safety. In essentially every case, the ecological consequences described in this chapter were discovered after the fact, by researchers investigating unexplained wildlife mortality or population decline, rather than anticipated in advance by whoever specified the lighting design.

This pattern — technology deployed for one purpose, with significant unintended ecological consequences only identified and studied well after widespread adoption — will be familiar from the historical chapters earlier in this book, and it recurs, in a very different but structurally similar form, in the discussion of satellite megaconstellations later in this book: a technology developed and deployed at enormous speed and scale, for reasons entirely unrelated to its effect on the night sky, with the full scope of that effect only becoming clear to the relevant scientific community after a very large fraction of the eventual damage was already, in effect, locked in.

The next chapter turns from these ecological consequences to a closely related but distinct question: what has this same transformation of the night done to human beings — not merely to our cultural relationship with the sky, as touched on in the previous chapter, but to our biology, our sleep, and our health.

Firefly decline: a familiar species, an underappreciated warning

Among the most emotionally resonant examples of insect-population impact from artificial light, at least for readers in regions where the phenomenon is culturally familiar, is the documented decline of firefly populations across parts of North America and Asia. Fireflies rely on precisely timed bioluminescent flash patterns, specific to each species, to locate and identify potential mates in the darkness of a summer evening – a communication system that depends entirely on sufficient ambient darkness for the comparatively faint bioluminescent signal to be visible and distinguishable against the background night.

Research examining firefly population trends in both wild and increasingly developed areas has identified artificial light at night as one of the more consistently implicated factors in observed local population declines, alongside habitat loss and pesticide use – a research finding consistent with the basic mechanism this chapter has described throughout: artificial light does not merely add visual clutter to a nocturnal species' environment, it can directly interfere with the specific behavioral signal, in this case a mating display, that the species' entire reproductive strategy depends on. Firefly tourism, which has grown considerably in recent years at a small number of well-known synchronous-flashing firefly sites (certain specific U.S. national park locations among them), has itself become an interesting minor case study in the broader dark-sky tourism economics discussed in Chapter 9, since visitors specifically seeking to witness undisturbed firefly displays have a direct practical interest in the same kind of shielded, minimal outdoor lighting that benefits stargazers, aligning two seemingly distinct conservation interests around the same underlying policy solution.

Urban bird song timing: a subtler behavioral shift

A further, somewhat less dramatic but well-documented example of artificial light's behavioral effects on wildlife involves the timing of bird song. Multiple field studies conducted in and around urban areas with significant artificial night lighting have found that certain songbird species begin their dawn chorus — the burst of vocal activity many bird species engage in around sunrise, believed to serve territorial and mate-attraction functions — measurably earlier in more heavily artificially lit locations compared to darker control sites, in some documented cases beginning well before actual sunrise in the most heavily lit urban environments, apparently because artificial light is disrupting the birds' internal perception of when dawn is actually approaching.

The downstream consequences of this shifted timing remain an active area of ongoing research, but several plausible concerns have been raised in the relevant literature: shifted song timing could affect the birds' own sleep and rest patterns, could alter competitive dynamics between individuals or species whose song timing shifts by different amounts in response to the same local lighting conditions, and in migratory species, could interact with the broader migratory timing and orientation disruptions discussed earlier in this chapter, compounding rather than existing independently from other documented light-pollution effects on the same populations. This example is worth including specifically because it illustrates that light pollution's ecological effects are not limited to the more dramatic, immediately visible cases — sea turtle disorientation, building-collision mortality — already discussed at length above, but extend into subtler behavioral and physiological shifts that require careful, sustained field research to even detect, let alone fully understand the downstream consequences of.

The scale problem: billions of fixtures, one mechanism

It is worth stepping back, at the close of this chapter, to emphasize a point that connects all of the specific examples discussed above: the underlying physical mechanism responsible for each of these documented ecological effects — an evolved behavioral or physiological response to natural light cues, disrupted by an evolutionarily novel artificial light source arriving on a timescale far too short for adaptation — is essentially the same mechanism in every case, whether the affected species is a sea turtle hatchling, a migratory warbler, a firefly, or a foraging bat. What varies between cases is simply which specific behavior (natal beach orientation, migratory navigation, mating display, foraging pattern) that shared underlying mechanism happens to disrupt for a given species.

This matters because it means the cumulative ecological cost of artificial light pollution, considered across the full range of affected species and behaviors this chapter has surveyed, is not confined to a small number of unusually vulnerable or unusually well-studied species. It represents a single, pervasive environmental change — the large-scale, rapid elimination of natural darkness across enormous fractions of the terrestrial and coastal environment — interacting with an correspondingly enormous number of independently evolved behavioral systems across the animal kingdom, each of which happens to have been calibrated, over evolutionary timescales, to a natural light-dark cycle that a very large fraction of the planet's land surface no longer reliably provides. The full scope of that cumulative effect, spread across the enormous number of individual species affected in ways both dramatic and subtle, almost certainly exceeds what the specific, well-documented case studies in this chapter can fully capture, simply because the resources and sustained field research required to document every affected species and behavior in comparable depth to, say, the sea turtle literature, do not remotely exist.

Fish, freshwater systems, and a less familiar impact category

Beyond the sea turtle, bird, insect, and amphibian examples already discussed at length in this chapter, it is worth briefly noting a less widely publicized but increasingly documented category of impact: freshwater and coastal fish species. Research examining artificial light effects on fish behavior has found documented disruption to a range of behaviors including nocturnal feeding patterns, predator-avoidance behavior, and, in some anadromous species (species that migrate between freshwater and marine environments as part of their life cycle, salmon being the best-known example), the timing of migratory movements, which in some species are cued in part by natural light-level changes that artificial lighting near rivers, dams, and coastal areas can measurably disrupt.

A specific and well-documented example involves outmigrating juvenile salmon, whose downstream migration timing and predator-avoidance behavior in some studied river systems has been shown to shift measurably in the presence of artificial lighting from bridges, riverside development, and dam infrastructure, potentially increasing predation risk during an already vulnerable life stage. While this research area remains less extensively developed than the sea turtle and migratory bird literature discussed earlier in this chapter, it illustrates the same underlying point already established repeatedly throughout this chapter: artificial light's behavioral disruption extends into ecological systems and species groups considerably beyond the small number of most famous, most frequently cited examples, and the true cumulative scope of ecological impact addressed by this chapter likely exceeds what current research, concentrated disproportionately on a small number of especially charismatic or especially convenient-to-study species, has yet been able to fully document.

Restoration success stories

It is worth closing this chapter's ecological discussion on a constructive note by briefly mentioning that targeted lighting interventions specifically designed around the mechanisms discussed in this chapter have, in a number of well-documented cases, produced measurable conservation benefit. The turtle-safe lighting ordinances mentioned earlier in the context of Florida nesting beaches represent one such success, with monitoring data from participating beaches generally showing improved hatchling sea-finding success following ordinance adoption and enforcement. Bird-collision "Lights Out" programs in cities along major migratory flyways, also mentioned earlier, have shown measurable reductions in both bird attraction and collision mortality specifically on nights when participating buildings actually honor their lighting-reduction commitments, providing some of the clearest available real-world evidence that the ecological harms documented throughout this chapter are not merely regrettable but essentially fixed costs of modern lighting, but are instead genuinely responsive to the kind of targeted, relatively modest intervention this book's final chapter will discuss at greater length and in a more general policy context.

Moths, pollination, and an underappreciated agricultural link

Beyond the general insect-decline concerns already discussed in this chapter, it is worth noting a more specific consequence for agriculture: many moth species, active primarily at night, serve as significant pollinators for a range of wild and cultivated plants, a pollination role considerably less publicly recognized than the daytime pollination services provided by bees, but ecologically significant in a number of documented plant communities. Research examining moth pollination activity near artificial light sources has found reduced pollination visitation rates in illuminated areas compared to darker control sites, plausibly linked to the same disorientation and attraction-to-light mechanisms discussed earlier in this chapter, raising the possibility, still being actively researched, that widespread artificial lighting may carry indirect agricultural and wild-plant-reproduction costs beyond the more direct wildlife-mortality concerns this chapter has emphasized, a genuinely underappreciated dimension of light pollution's ecological reach.

Zoos, aquariums, and managed-care lighting standards

A final, practically oriented example worth noting is the increasing attention paid to artificial lighting design within zoos and aquariums housing nocturnal or light-sensitive species, where professional animal-care standards have increasingly incorporated the same basic circadian and behavioral principles this chapter has discussed at length for wild populations. Modern nocturnal-house exhibits, for instance, commonly use reversed lighting cycles — dim during the institution's daytime visiting hours and brighter during actual nighttime hours when the facility is closed — specifically to allow diurnal human visitors to observe naturally nocturnal animals during their active period without exposing those animals to the kind of disruptive, mistimed light exposure this chapter has otherwise described as ecologically harmful, a small-scale, deliberately engineered counterexample to the largely unplanned, ad hoc lighting patterns responsible for most of the wild-population impacts discussed throughout this chapter.

Bats beyond the Luxor: a wider foraging pattern

Beyond the specific Luxor beam case study discussed at length earlier in this chapter, researchers studying bat foraging behavior more broadly have documented a consistent pattern across many species and locations: while some bat species readily exploit artificially lit areas as concentrated insect-foraging opportunities, in the manner described for the Luxor beam, other species appear to actively avoid brightly lit areas entirely, suggesting that artificial lighting restructures bat community composition in a given area by favoring light-tolerant species at the expense of light-averse ones, rather than uniformly benefiting or harming bats as a single ecological group.

Coastal development and combined lighting-noise stressors

Researchers studying coastal wildlife disruption have increasingly emphasized that artificial light rarely operates as an isolated stressor; coastal development bringing new lighting typically brings increased noise, human presence, and habitat modification simultaneously, and disentangling light pollution's specific independent contribution to a given documented wildlife impact from these compounding factors remains a genuine methodological challenge across much of the research this chapter has surveyed, one that the best-designed studies address through careful site-matched comparison but that inevitably leaves some residual uncertainty in any single study's specific attribution of harm to light alone.

Chapter 8 – The Human Night

Human beings are, in the strict biological sense, just as thoroughly shaped by the natural cycle of light and darkness as any of the species discussed in the preceding chapter. This chapter examines two distinct but related dimensions of that fact: the well-established physiological science of how artificial light at night affects human sleep and health, and the less rigorously measurable but arguably no less significant cultural and psychological consequences of losing routine access to a visible night sky.

The circadian system: an ancient clock

Nearly every cell in the human body operates according to an internally generated roughly-24-hour rhythm – the circadian rhythm – governed by a master clock located in a small region of the brain called the suprachiasmatic nucleus, which sits just above the point where the optic nerves cross and receives direct input from specialized light-sensitive cells in the retina. This system evolved, like the analogous systems described in the previous chapter for other species, over a vastly longer timescale than artificial lighting has existed, calibrated to a world in which the dominant and essentially only significant nighttime light source was, at most, moonlight and starlight, several orders of magnitude dimmer than even modest indoor electric lighting, let alone the direct glare of an outdoor floodlight or an illuminated screen held inches from the face.

The specific retinal cells responsible for feeding light information to this master clock — intrinsically photosensitive retinal ganglion cells, discovered and characterized only in the early 2000s, considerably later than most other basic components of human visual physiology — are most sensitive to blue-wavelength light, a detail with direct practical consequences discussed later in this chapter and in the solutions-focused Chapter 15. When these cells detect light, particularly blue-enriched light, during the biological night, they signal the suprachiasmatic nucleus to suppress production of melatonin, a hormone secreted by the pineal gland that plays a central role in regulating the timing of sleep onset and a range of other physiological processes tied to the day-night cycle. In a natural, pre-electric environment, this system worked exactly as intended: melatonin production rose reliably after sunset and fell before sunrise, providing a robust, consistent internal signal of nighttime that governed not just sleep timing but a wide array of other circadian-linked processes, including body temperature regulation, certain hormone release patterns, and aspects of immune and metabolic function.

Artificial light as a false signal

Artificial light at night disrupts this system by presenting the retina with a light signal that, in evolutionary terms, should not exist: bright, often blue-enriched illumination during the hours the body's circadian machinery expects darkness. A substantial body of clinical and laboratory research, accumulated particularly over the past two to three decades as researchers have had increasingly precise experimental tools for measuring melatonin levels and sleep architecture, has demonstrated that exposure to artificial light in the evening and nighttime hours measurably suppresses melatonin production, delays the timing of sleep onset, and can shift the overall phase of a person's circadian rhythm later relative to the external clock time — a phenomenon with obvious and well-documented consequences for next-day alertness, mood, and cognitive performance, and, with sustained chronic exposure, associations in the epidemiological literature with a considerably broader range of negative health outcomes.

This research base is most extensive and most rigorous with respect to indoor lighting and, especially, self-illuminated screens — smartphones, tablets, computer monitors, and televisions — held or viewed at close range in the hours before intended sleep, since these sources deliver a much higher effective light dose to the retina than most outdoor ambient light pollution, however bright a given city's skyglow might appear. It would be an overstatement to claim that outdoor light pollution alone, entering through a bedroom window, is responsible for the same scale of melatonin suppression as a smartphone screen held eighteen inches from the face; the physics of light intensity at the retina depend enormously on both source brightness and viewing distance, and most people's most intense nighttime light exposure by far comes from close-range screens and indoor lighting, not ambient outdoor skyglow. That said, a growing and increasingly rigorous body of research specifically addresses outdoor light pollution's contribution as well: multiple population-level studies have found statistically significant associations between higher levels of outdoor artificial light at night — measured using the same satellite-derived brightness data discussed in Chapter 6 — and increased rates of self-reported sleep disturbance, and a smaller number of studies have extended this to associations with broader health outcomes, including some research linking higher residential outdoor light pollution exposure to modestly elevated risk of certain metabolic and mood-related conditions, though as with most population-level epidemiological research in this area, establishing a clean causal pathway independent of other correlated urban environmental factors (noise, air pollution, socioeconomic variables) remains genuinely difficult, and researchers in this field are generally careful to describe these as documented associations warranting further investigation rather than settled, fully causally established findings.

Shift work as an instructive extreme case

A related and more rigorously established body of evidence comes from research on night-shift and rotating-shift workers, whose artificial light exposure and sleep-timing disruption is considerably more extreme, sustained, and easier to study directly than the generally more modest exposure represented by ambient residential light pollution. Multiple large-scale epidemiological studies of shift workers have found associations between long-term night-shift work and elevated risk of a range of health conditions, including certain metabolic disorders, cardiovascular disease, and, in some studies, certain cancers, findings serious enough that the World Health Organization's International Agency for Research on Cancer classified "shift work that involves circadian disruption" as a probable human carcinogen in 2007, a classification that has prompted a considerable amount of subsequent research attempting to isolate which specific mechanisms — sleep deprivation, meal-timing disruption, melatonin suppression, or some combination — are most responsible for these observed associations.

While shift work represents a far more extreme and sustained circadian disruption than typical exposure to ambient outdoor light pollution, this body of research is relevant to this chapter because it firmly establishes, in a context where the underlying mechanism (artificial light exposure at biologically inappropriate times, disrupting the circadian system) is not in serious scientific dispute, that circadian disruption of this general type carries genuine, measurable health consequences — providing a well-evidenced mechanistic foundation for the more preliminary but growing research suggesting that milder, chronic exposure to outdoor light pollution may carry smaller versions of similar risks for the broader population, not merely for night-shift workers specifically.

Beyond biology: the cultural and psychological loss

Chapter 1 of this book described, at some length, the extent to which the night sky functioned, across essentially every human culture prior to the twentieth century, as shared cultural infrastructure: a common visual and symbolic resource that supported navigation, timekeeping, agriculture, religion, mythology, and philosophy alike, available identically to any observer regardless of wealth or status. The near-total disappearance of naked-eye access to that resource for a large fraction of the world's population, documented in Chapter 6, represents a loss that is more difficult to quantify in the language of epidemiology or public health, but that a considerable body of writing — from professional astronomers, environmental philosophers, psychologists, and cultural historians alike — has argued is nonetheless real and significant.

Several distinct strands of argument recur in this literature. One emphasizes a straightforward loss of wonder and perspective: philosophers and scientists across many traditions have described direct encounters with a genuinely dark, star-filled sky as producing a distinctive psychological experience — sometimes described using the modern psychological term "awe," an emotional response associated in a growing body of research with measurable benefits including reduced stress, increased prosocial behavior, and a shift toward perceiving oneself as a smaller part of a larger whole (the "overview effect" reported by astronauts viewing Earth from space is frequently cited as a particularly intense version of a psychologically related phenomenon) — an experience that a night sky reduced to a handful of visible stars against a bright orange skyglow simply cannot reliably produce, regardless of how much a viewer might intellectually understand what they are missing.

A second strand of argument, more explicitly developmental and educational, concerns the effect of losing this experience specifically during childhood, a period when direct, sensory encounters with the natural world have been argued by developmental psychologists and educators to play a disproportionately significant role in shaping long-term environmental attitudes, scientific curiosity, and general engagement with the natural world — the general phenomenon sometimes described, in a related but distinct context concerning nature access more broadly, as "nature deficit disorder." Educators involved in astronomy outreach and dark-sky park programming frequently report, informally and through program evaluation data, that children experiencing a genuinely dark sky for the first time — commonly on a family trip to a national park or dark-sky preserve — display a visibly different and more intense reaction than could be produced by any photograph, planetarium show, or classroom lesson, however well designed; a number of science-education researchers have argued that this kind of direct, unmediated experience plays a role in fostering scientific interest that indirect representations cannot fully substitute for, though rigorous, controlled research isolating this specific effect from other confounding factors in a child's overall science education remains limited.

A third and more explicitly cultural strand of argument concerns what is lost at the level of collective human meaning-making rather than individual experience: the night sky's historical role, described at length in Chapter 1, as a shared symbolic resource underlying mythology, religious cosmology, and philosophical reflection across virtually every documented human culture. Some contemporary writers and dark-sky advocates have argued that a species that has effectively lost routine, visible access to the cosmological context in which nearly all of its foundational stories, calendars, and belief systems were originally developed has undergone a genuine, if difficult to measure precisely, cultural impoverishment — not merely an aesthetic loss, but the loss of easy, ongoing contact with the very phenomenon that generated an enormous fraction of human religious and philosophical tradition in the first place.

Holding the evidence at appropriate weight

It is worth being explicit, in closing this chapter, about the different levels of evidentiary confidence attached to the different claims it has surveyed. The physiological science of circadian disruption from artificial light, and its documented health associations, particularly in the shift-work literature, rests on a substantial and increasingly rigorous body of clinical and epidemiological research, even as researchers continue to refine understanding of exact mechanisms and dose-response relationships for milder, more typical exposure levels. The cultural and psychological arguments surveyed in the second half of this chapter are considerably harder to establish with equivalent rigor, resting more on qualitative observation, theoretical argument, and a smaller body of directly relevant empirical research than on the kind of large-scale, controlled studies available for the biological claims. Both categories of evidence are presented here not because they carry identical evidentiary weight, but because both are part of an honest accounting of what has actually been lost as the natural night has been engineered away — and because, taken together, they suggest that the consequences of that engineering extend well beyond the loss of a pleasant view, touching both human physiology and the deeper cultural and psychological fabric this book has traced back to its origins in Chapter 1.

The next several chapters turn from these costs toward what remains: the last genuinely dark places on Earth, the extraordinary scientific infrastructure built specifically to exploit what darkness survives, and — in the second half of this book — the new and rapidly escalating threat arriving not from the ground but from orbit.

Screens, bedrooms, and the modern sleep environment

While this chapter has been careful to distinguish outdoor ambient light pollution from the more intense, close-range light exposure delivered by smartphones, tablets, and other self-illuminated screens, it is worth examining the screen-exposure research in slightly more detail, both because it is the most rigorously established component of the broader circadian-disruption literature discussed in this chapter, and because it illustrates, in a more immediately actionable form, the same basic mechanism responsible for the outdoor light pollution effects that are this book's primary focus.

Laboratory studies measuring melatonin levels in participants exposed to bright screen light in the hours before scheduled sleep have consistently found measurable suppression of melatonin production compared to participants in dimmer or screen-free control conditions, with the degree of suppression generally correlating with both the brightness and the blue-light content of the specific screen and viewing conditions tested. Population-level survey research examining self-reported screen use before bed and various sleep-quality measures has found broadly consistent associations, with heavier pre-sleep screen use associated with delayed sleep onset, reduced total sleep time, and lower self-reported sleep quality across a considerable number of independent studies conducted in multiple countries and age groups, including a substantial body of research specifically focused on adolescents, a population of particular concern given both typically heavy device use and the specific developmental importance of adequate sleep during this life stage.

This body of research has prompted a range of practical recommendations from sleep-medicine professional organizations, including reduced screen use in the hour or more before intended sleep, use of device-level blue-light-reduction features (now built into most major smartphone and tablet operating systems, generally shifting screen color temperature warmer during evening hours), and, more broadly, maintaining a dark or minimally lit bedroom environment during sleep itself, a recommendation directly relevant to this chapter's discussion of outdoor light pollution, since bedroom light contamination from external streetlights, neighboring building lighting, or other outdoor sources entering through windows represents exactly the kind of avoidable additional nighttime light exposure this research suggests is worth minimizing, alongside the more heavily studied screen-exposure question.

Seasonal affective patterns and the broader question of light's role in mood

A further, more preliminary but genuinely relevant strand of research worth noting concerns the broader relationship between light exposure patterns and mood regulation. Seasonal affective disorder — a well-established pattern of depressive symptoms occurring predictably during specific seasons, most commonly winter in higher-latitude regions with substantially reduced daylight hours — is among the clearest established examples of light exposure directly affecting human mood and mental health at a population level, and its standard clinical treatment, bright-light therapy administered during specific times of day, works by directly manipulating a patient's light exposure pattern to correct an underlying circadian and mood-regulation disruption.

While seasonal affective disorder itself concerns insufficient daytime light exposure rather than excessive nighttime exposure, and is therefore not a direct light-pollution effect in the sense this book has otherwise used the term, its well-established existence as a genuine, clinically significant condition demonstrates clearly that light exposure timing and pattern have real, measurable effects on human mood regulation through mechanisms closely related to the circadian system discussed throughout this chapter. This established relationship has prompted a smaller but growing body of research specifically examining whether excessive or poorly timed nighttime light exposure — the outdoor light pollution and screen exposure discussed above — might carry analogous, if generally milder, mood-related consequences of its own, a research question that remains considerably less settled than the more directly established sleep-disruption findings discussed earlier in this chapter, but that a number of researchers in the field consider a genuinely promising and important area for continued investigation.

What a return to darker skies might offer, tentatively

It is worth closing this chapter's biological discussion with an appropriately cautious note about the flip side of the research surveyed above: if artificial light at night measurably disrupts sleep, circadian regulation, and potentially broader aspects of mood and health, it follows, at least in principle, that reducing unnecessary nighttime light exposure — both the close-range screen exposure most rigorously studied, and the more diffuse outdoor light pollution this book's chapters on solutions will address directly — could offer at least modest corresponding benefits. A small but growing body of intervention research, testing structured reductions in evening light exposure against various sleep and mood outcome measures, has generally found results consistent with this expectation, though the magnitude of benefit, unsurprisingly, appears to depend considerably on the severity of light exposure being reduced and the specific population studied.

This chapter has tried throughout to hold its claims to an appropriate and honestly stated level of evidentiary confidence, and that same caution applies here: this is not a claim that reducing outdoor light pollution specifically would produce dramatic, clinically significant health benefits on its own, a claim the current evidence does not clearly support. It is a more modest claim, but one worth stating plainly nonetheless: the biological case for treating unnecessary nighttime light exposure, in all of its forms, as a genuine if generally modest health consideration, rather than a purely aesthetic or environmental one, rests on a solid and growing body of research, and provides yet another concrete, evidence-based reason, alongside the ecological and cultural arguments developed throughout this chapter and the one preceding it, for taking the solutions discussed in Chapter 15 seriously.

Occupational exposure beyond classic shift work

Beyond the classic night-shift and rotating-shift-work populations discussed earlier in this chapter, it is worth noting that a considerably broader range of contemporary occupations now involve significant nighttime artificial light exposure than the traditional factory-shift-work model most extensively studied in the older circadian-disruption literature. Long-haul transportation workers, emergency responders, hospitality and security personnel, and a growing number of remote and digitally connected knowledge workers whose work schedules span multiple time zones or extend well into conventional nighttime hours all experience some version of the circadian disruption this chapter has discussed, even when their specific occupational category has received less dedicated research attention than the more traditional shift-work populations that formed the evidentiary basis for the International Agency for Research on Cancer's 2007 classification discussed earlier.

This broadening of who is actually affected by artificial-light-driven circadian disruption matters for how this chapter's findings should be understood: the health associations discussed earlier in this chapter are not a narrow concern relevant only to a shrinking population of traditional factory shift workers, but a broader public-health consideration relevant to an increasingly large and occupationally diverse fraction of the contemporary workforce, whose collective nighttime light exposure patterns look considerably different, and in aggregate considerably more extensive, than the mid-twentieth-century occupational patterns the foundational shift-work research was originally designed to study.

Children, adolescents, and developmental sensitivity

A further population-specific consideration worth addressing directly is the particular sensitivity of children and adolescents to the circadian and sleep effects discussed throughout this chapter. Adolescent sleep-timing research has established that the circadian rhythm naturally shifts later during adolescent development, a well-documented biological phenomenon behind widespread public-health advocacy for later school start times in a number of countries, and a number of researchers examining this population specifically have raised concern that the combination of this natural developmental shift with heavy evening screen use and, more relevant to this book's central focus, ambient nighttime light pollution, may compound to produce sleep-timing disruption more severe than either factor would produce in isolation. This concern connects directly back to the cultural and educational arguments developed in the second half of this chapter regarding children's access to genuinely dark night skies, suggesting that the loss of that access may carry compounding costs for the specific population of children and adolescents that extend from the more clearly established biological sleep-science findings through to the more preliminary cultural and developmental arguments this chapter surveyed toward its close.

Hospital and healthcare lighting environments

A final, practically significant application of the circadian research discussed throughout this chapter concerns hospital and long-term care facility lighting design specifically. A growing body of research and design guidance has argued that traditional hospital lighting, often maintained at fairly constant, moderately bright levels around the clock for staff convenience and patient monitoring purposes, may inadvertently work against patient recovery by disrupting exactly the circadian regulation this chapter has described, and a number of newer hospital facilities and renovation projects have begun incorporating circadian-informed lighting design specifically intended to reinforce, rather than disrupt, patients' natural day-night rhythm during their stay, an application of this chapter's underlying science with direct, practical relevance to patient outcomes rather than the more diffuse population-level associations this chapter has otherwise discussed.

Cross-cultural variation in nighttime light norms

It is worth noting briefly that cultural attitudes toward nighttime brightness, and correspondingly toward the health and psychological considerations discussed throughout this chapter, vary meaningfully across different societies, shaped by differing historical relationships with safety, commerce, and communal nighttime activity. Some cultures have long-standing traditions of vibrant, illuminated nighttime social and commercial life extending well past sunset, while others have historically maintained more clearly demarcated boundaries between daytime activity and a comparatively dark, quiet nighttime period. These cultural variations complicate any simple, universal policy prescription regarding acceptable nighttime lighting levels, suggesting that the solutions discussed in Chapter 15, however well-evidenced in the specific contexts they have been studied, likely need some degree of local cultural adaptation to be both effective and broadly acceptable across the genuinely diverse range of societies this book's global-scale concerns ultimately touch.

Athletic performance and evening light exposure

A further, narrower strand of circadian research worth noting examines the relationship between evening artificial light exposure and next-day athletic and cognitive performance, with several sports-science studies finding that athletes exposed to bright, blue-enriched lighting in the hours before sleep show measurably reduced sleep quality and next-day reaction-time performance compared to athletes following dimmer, warmer-toned evening lighting protocols, a specific, practically oriented illustration of the broader circadian-disruption mechanisms this chapter has described in more general population-health terms.

Elderly populations and heightened circadian sensitivity

Research on aging and circadian regulation has found that older adults generally experience reduced melatonin production and a naturally more fragile circadian rhythm compared to younger adults, suggesting that this population may be disproportionately sensitive to the nighttime light disruption discussed throughout this chapter, a consideration increasingly reflected in specialized lighting guidance for senior living facilities, which have begun adopting circadian-informed lighting design broadly similar to the hospital-lighting approaches discussed earlier in this chapter.

Chapter 9 — The Last Places on Earth

If the preceding chapters have traced a story of near-total loss, this chapter pauses on what survives, and on the deliberate efforts, still relatively young as organized movements go, to identify, protect, and in some cases restore genuinely dark night skies. It closes with an extended look at one of the most scientifically and symbolically important dark places remaining on the planet — the Atacama Desert of northern Chile — and the uncomfortable discovery that even there, "far enough" may not be a permanent condition.

The rise of the dark-sky movement

Organized concern about light pollution, as distinct from isolated complaints by individual astronomers, emerged as a coherent movement principally from within the amateur and professional astronomical community beginning in the 1970s and 1980s, as the deteriorating observing conditions described in Chapter 5 became impossible for working astronomers to ignore. The International Dark-Sky Association, founded in 1988 by astronomers David Crawford and Tim Hunter, became the most influential organization in this space, developing both a certification program for parks, reserves, and communities that meet specific outdoor lighting and sky-darkness standards, and a broader advocacy and public-education program aimed at promoting what the organization terms "responsible outdoor lighting" — an approach, discussed at greater length in Chapter 15, that generally does not argue for eliminating outdoor lighting but rather for using it more efficiently: shielding fixtures so light is directed downward where it is needed rather than wastefully upward, using appropriate brightness levels rather than defaulting to maximum output, and employing timers or motion sensors so that lighting is not left running continuously in locations where it serves no purpose during quiet overnight hours.

The organization's certification categories — International Dark Sky Parks, Reserves, Sanctuaries, Communities, and Urban Night Sky Places, each with somewhat different criteria reflecting different types of locations and levels of achievable darkness — have grown from a small handful of initial certifications in the early 2000s to several hundred certified locations across dozens of countries by the 2020s, spanning everything from remote wilderness reserves to entire towns that have adopted comprehensive dark-sky-friendly lighting ordinances. This growth reflects both increasing public awareness of light pollution as an issue in its own right and a growing recognition, discussed further below, that dark skies carry real economic value as a tourism draw, giving certified communities a practical incentive beyond pure environmental stewardship to pursue and maintain certification.

National parks as accidental refuges

Beyond formal dark-sky certification, a considerable amount of the remaining dark sky accessible to ordinary people, particularly in North America, exists within national parks and other large protected wilderness areas, which frequently retain dark-sky conditions as an incidental consequence of restrictions on development and permanent settlement within their boundaries, rather than through lighting policy specifically designed with astronomy in mind. The

United States National Park Service has, since the early 2010s, developed dedicated night-sky programming and monitoring at many major parks, using standardized sky-quality measurement equipment to establish baseline darkness readings, hosting public astronomy and night-sky interpretation programs, and, in a growing number of cases, actively pursuing International Dark Sky Park certification as an explicit management goal rather than treating dark skies as a mere byproduct of the park's primary conservation mission.

This shift matters because it represents an evolution in how dark skies are conceived institutionally: from an unplanned, incidental benefit of wilderness protection toward an explicitly recognized and actively managed resource in its own right, worthy of the same kind of deliberate stewardship long applied to a park's wildlife, water, or geological features. Several major national parks — including a number in the American Southwest, whose combination of high elevation, low regional humidity, and relatively sparse surrounding development make them particularly well suited to dark-sky preservation — have become significant destinations specifically for stargazing tourism, with ranger-led night-sky programs regularly drawing crowds comparable to a park's more traditional daytime attractions.

Dark skies as economic asset

The economic argument for dark-sky preservation, largely absent from the earlier, purely scientific and conservation-focused phase of the dark-sky movement, has become an increasingly significant part of the case made to municipal governments and tourism boards for adopting dark-sky-friendly lighting policy. Communities that have pursued and achieved International Dark Sky Community status, or that market themselves more informally as stargazing destinations, have in a number of documented cases reported measurable increases in tourism specifically tied to that reputation — visitors traveling explicitly for astrophotography, stargazing tours, and the simple novel experience, for visitors from heavily light-polluted home regions, of seeing a genuinely dark sky for the first time in their adult lives.

This economic dimension gives dark-sky preservation a rare quality among environmental protection arguments: rather than requiring a purely values-based tradeoff between economic development and environmental preservation, it offers, at least for communities with suitable existing darkness and reasonable proximity to population centers capable of generating tourist traffic, a case in which preserving and even actively marketing natural darkness can align directly with local economic interest, potentially making dark-sky-friendly lighting ordinances considerably easier to enact politically than environmental regulations that impose a straightforward cost with no comparably direct local financial benefit.

The Atacama: the paradigmatic dark place

No discussion of the world's remaining dark places would be complete without extended attention to the Atacama Desert, a narrow strip of land running along the Pacific coast of northern Chile between the Andes and the coastal mountain range, and one of the driest, most consistently clear, and most astronomically valuable locations on the entire planet. The desert's combination of extremely low humidity (parts of the Atacama rank among the driest places on Earth by any measure, with some weather stations having recorded no measurable rainfall across periods of years), high elevation at many of its key astronomical sites, minimal cloud cover across most of the year, and, critically for this book's purposes, extremely low population density and correspondingly minimal artificial light pollution across vast stretches of the desert, has made it one of the most sought-after locations on Earth for ground-based astronomical observation.

The scientific infrastructure built to exploit these conditions is considerable: the European Southern Observatory operates several major astronomical facilities in the Atacama region, including the Very Large Telescope at Cerro Paranal, among the most scientifically productive ground-based optical observatories in the world, and the Atacama Large Millimeter/submillimeter Array, a large radio-telescope facility situated on the high Chajnantor plateau. Additional major facilities operated by other international consortia are sited elsewhere in the same general region, drawn by the same combination of atmospheric and darkness conditions. Collectively, these facilities represent billions of dollars of accumulated international scientific investment, concentrated in this specific region precisely because of qualities — darkness, dryness, atmospheric stability — that are, as this book's overall argument emphasizes, neither permanent nor immune to encroachment.

"Surely we're far enough now" — and then, not quite

The Atacama functions in this book's narrative as the clearest available test case for a question that runs through several of the preceding chapters: is there anywhere left that is simply, permanently far enough from encroaching development to remain reliably dark? For decades, the honest answer for the Atacama specifically was close to yes — remote, sparsely populated, protected in practice by its sheer inhospitability to large-scale human settlement, the desert's darkness seemed considerably more secure than that of, say, a national park within easy driving distance of a major metropolitan area.

That assumption has come under genuine and specific pressure in recent years. Industrial and energy development projects proposed for construction near some of the region's most scientifically important observatory sites — including, in reporting that drew considerable international attention within the astronomical community, a large proposed industrial energy project near the Paranal region specifically — have raised direct concerns among the astronomical organizations operating major facilities there about potential light pollution and other environmental impacts (including dust and vibration, which can degrade telescope performance independent of sky brightness) from nearby development, even in a location whose remoteness had long been treated as its primary and most durable protection. The specific dispute over development near Paranal became prominent enough to draw international press coverage and formal objections from the scientific institutions operating observatories in the area, precisely because it represented, in concrete and immediate form, the uncomfortable realization that even one of the most remote and inhospitable dark-sky locations on the planet is not automatically protected from the same pressures — economic development, energy infrastructure, population growth — that have already consumed dark skies everywhere else on Earth that lacked either extreme remoteness or deliberate legal protection.

This is, in miniature, the central tension this book keeps returning to in different forms: darkness that seems secure because of its remoteness or inhospitability is not actually protected by any inherent, durable property of the location itself; it is protected, when it is protected at all, only by the ongoing absence of sufficiently motivated development pressure, and that absence cannot be assumed to be permanent anywhere, including in one of the driest and most remote deserts on the planet, specifically chosen by international astronomical consortia because it seemed, for decades, about as far from encroaching civilization as a location on Earth's surface could plausibly be.

What "the last places on Earth" actually means

The chapter's title is deliberately double-edged. In one sense, it refers straightforwardly to the specific locations — certified dark-sky parks and reserves, protected wilderness areas, remote deserts like the Atacama — that currently retain something close to the pristine, Class 1 or Class 2 sky conditions described in Chapter 1, and that a determined traveler can still, with sufficient effort and expense, reach today. In a more unsettling sense, though, it is a genuine open question whether these currently-dark locations will remain the last places on Earth where such skies survive, or whether the pattern already visible at Paranal — encroaching development even in the most remote and seemingly secure dark-sky locations — will eventually catch up with all of them, leaving an even smaller, more fragile, and more heavily protected set of true dark-sky refuges than currently exist.

The next chapter turns from this question of where darkness survives to a related and equally important question: what, precisely, that surviving darkness is used for — the extraordinary scientific instruments built specifically to exploit it, the engineering and economics behind them, and the specific reasons that darkness itself, rather than merely a large telescope, remains such an irreplaceable scientific resource.

How dark-sky certification actually works

It is worth describing, in a bit more procedural detail, what achieving International Dark Sky certification actually involves for a park, reserve, or community, since the process itself illustrates both the seriousness with which the movement approaches the goal and the practical challenges involved in reversing, or at least halting, existing light pollution at a given site.

An applicant location — whether a national or state park, a private wilderness reserve, or an entire town or city government — must typically submit sustained sky-brightness measurement data, collected using calibrated sky quality meters at multiple points across the site over an extended period, demonstrating that the location meets specific quantitative darkness thresholds appropriate to its certification category (the thresholds differ meaningfully between, for instance, a large, sparsely developed wilderness reserve and a certified "Urban Night Sky Place," which by definition permits considerably more residual light pollution given its urban or suburban context, but still requires documented, deliberate efforts to minimize unnecessary lighting relative to comparable non-certified locations). Beyond the measurement requirement, applicants must typically demonstrate an active outdoor lighting management plan: documented policies governing new construction, fixture replacement, and, in the case of full municipal certification, a genuine local ordinance with enforcement mechanisms, rather than a purely voluntary or aspirational commitment. Certification, once achieved, is not permanent; certified locations are subject to periodic review and can, in principle, lose certification if lighting conditions deteriorate and are not corrected, a structure explicitly designed to incentivize sustained, ongoing management rather than a one-time compliance exercise.

This structure — measurable thresholds, a genuine management plan, and periodic re-review — has proven influential well beyond the specific certified locations themselves, since a considerable number of municipalities that have never pursued formal certification have nonetheless adopted lighting ordinances modeled closely on the technical standards developed through the certification process, using the International Dark-Sky Association's published model lighting ordinance language as a starting template for their own local regulations, a pattern of policy diffusion that has meaningfully extended the practical influence of the dark-sky movement considerably beyond the specific set of formally certified sites.

Chile's broader dark-sky protections

Beyond the specific Paranal dispute discussed later in this chapter, it is worth noting that Chile as a whole has developed some of the most comprehensive national-level light pollution regulation of any country in the world, a policy commitment directly traceable to the country's outsized role, described throughout this chapter, in hosting major international astronomical infrastructure. Chilean national environmental regulation includes specific outdoor lighting standards for areas near major observatory sites, restricting fixture types, permitted light output, and shielding requirements for both public and private lighting within defined zones surrounding the country's most scientifically significant sites, a regulatory framework developed through sustained collaboration between the Chilean government and the international astronomical organizations operating within its borders.

This national-level regulatory commitment illustrates a policy pathway distinct from, and arguably more robust than, the municipal-by-municipal ordinance approach exemplified by Flagstaff and discussed at length in Chapter 15: a national government, recognizing dark skies as a strategically significant scientific and economic resource (given the substantial international investment and ongoing operational spending associated with hosting major observatories), can in principle establish binding protections at a scale no single municipality could achieve unilaterally. Whether this national-level regulatory framework proves sufficient to withstand the specific development pressures described below in the Paranal case remains, as this chapter's closing sections make clear, a genuinely open and closely watched question, both within Chile and among the broader international astronomical community with a direct stake in the outcome.

Dark-sky preserves beyond North America and Chile

While this chapter's primary case studies focus on North American national parks, Flagstaff-style municipal ordinances, and the Chilean observatory sites, it is worth noting briefly that the dark-sky preservation movement has become genuinely global in scope. Certified dark-sky reserves and parks now exist across Europe (including several sites in the United Kingdom, Germany, and Scandinavia, regions that, as noted in the previous chapter, otherwise rank among the more heavily light-polluted parts of the world in aggregate, making their certified dark-sky enclaves particularly valuable as accessible refuges for a large regional population with limited realistic access to more remote dark-sky locations), in Namibia (home to one of the highest-profile dark-sky reserves in Africa, benefiting from the country's exceptionally low population density and correspondingly minimal regional light pollution), and in Australia, whose vast, sparsely populated interior offers some of the most extensive naturally dark sky remaining anywhere on the planet, a substantial fraction of it entirely without any formal certification simply because so little of it faces any immediate development pressure requiring the kind of deliberate protective designation more urgently needed in densely populated regions elsewhere.

This global spread matters for the argument of this book because it demonstrates that the impulse to identify, measure, and protect dark skies deliberately, as a resource worth the same kind of conscious stewardship traditionally applied to other natural resources, is not a narrow or geographically parochial concern limited to a small number of wealthy, astronomically inclined countries, but has become a genuinely international movement, responding to what is, as the preceding chapters of this book have documented at length, a genuinely global phenomenon.

Astro-tourism as a growing travel category

Beyond the specific municipal and park-level economic benefits discussed earlier in this chapter, it is worth noting that dark-sky travel has emerged as a recognized and growing subcategory within the broader travel and tourism industry over the past decade or so, with dedicated tour operators, specialized astrophotography workshops, and destination marketing campaigns increasingly built explicitly around access to genuinely dark night skies as a primary attraction rather than an incidental amenity. Industry analyses and tourism-board reporting from several dark-sky-designated regions have documented specific visitor-spending figures attributable to astro-tourism activity, and a number of remote regions with historically limited conventional tourism appeal — sparsely populated desert and high-altitude areas that might otherwise struggle to attract visitors on the basis of more conventional attractions — have found dark-sky tourism marketing to be a genuinely viable economic development strategy specifically because their remoteness and low population density, disadvantages for most other forms of tourism development, are direct advantages for this specific niche.

This growth has, in turn, created a reinforcing economic incentive for the kind of lighting ordinances and certification efforts discussed throughout this chapter: a community whose local economy has come to depend, even partially, on its dark-sky reputation has a direct and ongoing financial stake in maintaining the darkness that reputation depends on, providing a form of institutional accountability for sustained lighting-policy commitment that goes beyond the purely values-based motivation that might otherwise be the only force sustaining such policy over multiple decades and multiple changes in local government leadership.

Beyond Paranal: a broader pattern of encroachment risk

While this chapter has focused at length on the Paranal development dispute as its central case study in encroachment risk even at the most remote and seemingly secure dark-sky sites, it is worth noting that comparable, if generally less internationally publicized, encroachment concerns have been raised by astronomical institutions operating at other major remote observatory sites as well, generally involving some combination of regional population growth, mining or energy infrastructure development, and expanding transportation networks bringing artificial lighting closer to previously isolated facilities than had been the case when those facilities were originally sited decades earlier. This broader pattern reinforces the central point this chapter has tried to establish: Paranal is a particularly well-documented and high-profile example of a more general dynamic, not an isolated or unusual case, and the underlying pressures driving it — population growth, economic development, and expanding infrastructure, none of which has any inherent regard for astronomical darkness requirements — apply in some degree to nearly every remote observatory site currently in operation anywhere in the world, making sustained, active protective policy, of the kind discussed throughout this chapter, a continuing necessity rather than a one-time achievement at any dark-sky site, however remote.

Volunteer sky-quality monitoring networks

Beyond the formal certification processes discussed earlier in this chapter, a considerable amount of ongoing dark-sky monitoring is conducted by volunteer citizen-scientist networks, coordinating with programs such as Globe at Night, discussed in Chapter 6, to collect the sustained, geographically distributed measurement data that both certification review and broader trend research depend on. These volunteer networks represent a low-cost but genuinely valuable complement to the more resource-intensive professional monitoring conducted at major observatory sites, extending useful data collection to a much larger number of locations than professional researchers alone could practically monitor, and providing an accessible, low-barrier way for members of the public to contribute directly and meaningfully to the broader dark-sky research and advocacy effort this chapter has described.

Indigenous land stewardship and dark-sky overlap

In a number of documented cases, dark-sky preservation goals have found productive alignment with indigenous land stewardship and cultural preservation efforts, since many indigenous communities maintain traditional territories in comparatively remote, sparsely developed regions that also happen to retain some of the darkest skies discussed throughout this chapter, and since a number of these communities maintain their own long-standing cultural and astronomical traditions, of the kind discussed in Chapter 1, directly tied to unobstructed access to the night sky. Several dark-sky reserve designations in different parts of the world have been developed in direct partnership with indigenous communities, incorporating traditional ecological and astronomical knowledge alongside the more conventional scientific criteria this chapter has otherwise emphasized, offering a promising model for dark-sky preservation that serves cultural and scientific goals simultaneously rather than treating them as separate, unrelated concerns.

Private land conservation easements

Beyond formal park and municipal designations, a number of dark-sky preservation efforts have relied on private land conservation easements, in which a landowner voluntarily and permanently restricts future development or lighting intensity on their property in exchange for tax benefits or simply out of personal commitment to conservation, a mechanism borrowed directly from the broader land-trust conservation movement and increasingly applied specifically to dark-sky preservation in regions where formal public park designation is not practical or available.

Astro-tourism infrastructure beyond lodging

Beyond lodging and guided tours, a number of dark-sky destinations have invested in dedicated observation infrastructure specifically for visiting tourists - permanent viewing platforms, telescope-rental programs, and interpretive signage explaining what visitors are seeing - representing a further, tangible economic and institutional commitment beyond lighting policy alone, reinforcing the practical, on-the-ground seriousness with which a number of these destinations have approached dark-sky tourism as a genuine, sustained economic development strategy rather than a passing marketing trend.

Cross-border reserve networks

A small number of dark-sky reserves span international borders, requiring coordinated lighting policy between neighboring national or provincial governments, an administratively complex but increasingly pursued model that offers a small-scale preview of the kind of international coordination challenge, discussed at much greater length in Chapters 11 through 13, that the orbital half of this book's argument raises on a dramatically larger and more consequential scale.

Chapter 10 — The Mountain Where We Look Back in Time

Why does a research telescope need to be on a remote, dark mountaintop at all? The question is worth asking directly, because the answer explains both why so much scientific and financial investment has been poured into the specific locations discussed in the previous chapter, and why the threats to those locations' darkness — whether from ground-based development or, as later chapters explore, from orbiting satellites — represent genuine threats to scientific capability, not merely aesthetic inconvenience.

Site selection: an exacting science

Modern ground-based observatory site selection is itself a specialized field, and the criteria involved go well beyond simple darkness. Altitude matters because a large fraction of the atmosphere's total mass, and a correspondingly large fraction of its water vapor, light-scattering aerosols, and turbulence, lies in the lowest few kilometers; a telescope situated at four or five thousand meters above sea level, as many of the world's premier observatory sites are, observes through dramatically less obstructing atmosphere than one at sea level, improving both the sharpness of images and the range of wavelengths (particularly in the infrared, heavily absorbed by atmospheric water vapor) that can be usefully observed at all.

Atmospheric stability — what astronomers call "seeing" — is a second, closely related criterion, referring to the degree to which turbulent air currents blur and distort incoming starlight before it reaches the telescope's mirror, the same physical phenomenon responsible for the twinkling of stars visible to the naked eye. Sites with unusually stable atmospheric conditions, generally associated with specific combinations of altitude, prevailing wind patterns, and distance from disruptive local terrain features, produce measurably sharper images even before accounting for sky brightness at all, and site-selection surveys for major new observatories typically involve years of direct atmospheric monitoring at candidate locations before a final site is chosen.

Aridity matters for multiple reasons beyond simply reducing cloud cover: dry air holds less water vapor to scatter and absorb light, and drier climates are also generally associated with fewer nights lost to cloud cover altogether, a critical practical consideration given the enormous cost, discussed below, of building and operating a major observatory — a facility that spends a third of its nights unable to observe due to weather represents a substantial waste of an extremely expensive capital investment. And, finally, of course, darkness itself — freedom from nearby artificial skyglow — remains an essential criterion, for exactly the reasons detailed throughout the earlier chapters of this book: even a small amount of artificial sky brightening measurably reduces the faintest, most scientifically interesting objects a telescope of a given size can detect, since detecting a faint astronomical signal fundamentally depends on distinguishing it from the background sky brightness behind it.

Sites that meet the standard

A relatively small number of locations on Earth satisfy all of these criteria simultaneously well enough to host the world's most scientifically productive major observatories, and it is not a coincidence that these sites cluster in a handful of specific regions: the Atacama Desert of northern Chile, discussed at length in the previous chapter, hosts an unusually large concentration of major international facilities for precisely the combination of reasons just described. High-altitude, arid volcanic locations in Hawaii have historically hosted a comparable concentration of major optical and infrared telescopes, similarly selected for altitude, atmospheric stability, and relative isolation, though Hawaiian observatory development has also become the subject of significant, ongoing debate involving the cultural and religious significance of these sites to Native Hawaiian communities — an important dimension of observatory siting controversy that operates independently of, but sometimes intersects with, the pure astronomical site-quality criteria discussed in this chapter. The Canary Islands, off the northwest coast of Africa, host another significant concentration of major telescopes, benefiting from a comparable combination of altitude and atmospheric stability at several specific island peaks. A handful of other sites around the world — in South Africa, Australia, and elsewhere — host additional major facilities, generally selected for some subset of the same core criteria adapted to regional geography.

The economics of a billion-dollar telescope

The financial scale of major modern observatories is worth stating explicitly, because it clarifies exactly what is at stake when a site's darkness or atmospheric quality is threatened by nearby development. Extremely large ground-based telescopes under construction or recently completed — instruments with primary mirrors twenty to forty meters in diameter, built from many smaller segments precisely aligned and continuously adjusted by sophisticated computer-controlled support systems — represent capital investments typically running well into the billions of dollars once construction, instrumentation, and the specialized infrastructure (access roads, housing for staff, power supply, and in many cases desalination or water-transport infrastructure for facilities in extremely arid regions) required to operate a research facility in a remote location are all accounted for. These costs are typically shared among international consortia of governments, universities, and research institutions specifically because no single national funding agency can easily bear the full cost alone, a funding structure that itself creates significant institutional inertia around site selection, since renegotiating an international partnership around a different site, once substantial construction has begun, is logistically and politically extremely difficult.

Given this scale of investment, the practical question a site's darkness answers is not abstract: for every increment of artificial sky brightening at a given observatory site, the faintest, most distant, most scientifically interesting objects that instrument can detect become measurably harder to distinguish from the background sky, directly reducing the scientific return on an investment of this magnitude. A doubling of local sky brightness at an observatory site does not merely make a slightly less pretty view for visiting VIPs; it can meaningfully reduce the maximum distance, and therefore the maximum look-back time into cosmic history, at which the facility's most demanding scientific programs can successfully detect their faintest targets — an effect equivalent, in some cases, to having built a smaller and less capable telescope than the one actually constructed, despite the full cost having already been paid.

What darkness actually buys: looking back in time

The chapter's title refers to a genuinely remarkable fact about astronomical observation that is easy to state but sometimes underappreciated even by otherwise well-informed readers: because light travels at a finite speed, and because the universe is so large that even that finite speed translates into enormous travel times across cosmic distances, looking at a sufficiently distant object is, quite literally, looking back in time — observing that object not as it exists today, but as it existed however many years ago the light left it on its journey to the telescope's mirror.

The most scientifically ambitious observational astronomy programs are specifically designed to push this look-back capability as far as possible: to detect the faintest, most distant galaxies, whose light has been traveling for the majority of the universe's roughly 13.8-billion-year history, in order to study the earliest phases of cosmic structure formation directly, rather than merely inferring them from theoretical models. Every increment of additional darkness at an observing site, every reduction in atmospheric turbulence, and every improvement in telescope mirror size and instrument sensitivity contributes, in combination, to pushing that observable frontier a little further back toward the earliest moments of the universe's history — which is precisely why the international astronomical community has continued to invest such extraordinary sums in identifying and protecting the darkest, most atmospherically stable sites remaining on Earth, and why threats to those sites, whether from nearby ground development of the kind discussed in the previous chapter, or from the satellite megaconstellations discussed beginning in the next chapter, are taken so seriously by the scientists and institutions who depend on them.

A resource unlike most others

It is worth closing this chapter by naming clearly what kind of resource sky darkness actually is, because the framing matters for the remainder of this book's argument. Sky darkness is not consumed in the way a mineral resource is consumed — using it does not deplete it directly. But it is finite in practice, in the specific sense that only a limited number of locations on Earth's surface satisfy the combination of altitude, aridity, atmospheric stability, and darkness required for premier astronomical observation, and every one of those locations is subject to gradual erosion from nearby development, exactly as the previous chapter's discussion of the Atacama demonstrated. It is also, unlike many other environmental resources, largely invisible to the people whose activities erode it: a distant industrial project, an expanding regional power grid, or a new highway interchange built many kilometers from an observatory site may contribute measurably to that site's sky brightness without anyone directly involved in approving or building that project ever considering astronomical darkness as a relevant factor in their decision at all.

This combination — genuinely finite, gradually eroding, and largely invisible to the parties responsible for eroding it — is precisely the profile of resource most likely to be gradually consumed without anyone making a deliberate decision to consume it, simply through the accumulated, uncoordinated pressure of many separate actors each pursuing entirely unrelated goals. It is a profile this book will return to explicitly in Chapter 14, once the second major thread of this story — light pollution arriving not from the ground but from orbit — has been fully introduced, because the same basic dynamic, playing out on a much faster timescale and with a very different set of responsible parties, turns out to describe the satellite megaconstellation problem just as precisely as it describes the terrestrial development pressure already threatening sites like Paranal.

That orbital story begins in the next chapter.

The instrument itself: mirrors, segments, and adaptive correction

It is worth describing, briefly, how the largest contemporary ground-based telescopes actually achieve the extraordinary light-gathering power and image sharpness referenced elsewhere in this chapter, since the engineering involved is directly relevant to understanding both the cost figures already discussed and the stakes involved when a site's atmospheric or darkness conditions are degraded.

Building a single, rigid primary mirror much beyond about eight meters in diameter using traditional glass-casting and grinding techniques becomes prohibitively difficult and expensive, since a sufficiently large single piece of glass becomes unacceptably heavy and prone to distorting under its own weight as the telescope is repositioned to track different parts of the sky through the night. The current generation of extremely large telescopes instead uses segmented mirror designs: dozens of smaller, precisely shaped hexagonal mirror segments fitted together to form a single much larger effective optical surface, each segment individually mounted on a computer-controlled actuator system capable of adjusting that segment's position and orientation many times per second to keep the entire assembly working together as a single, coherently focused optical surface despite temperature changes, wind, and the mechanical stresses of repositioning the telescope throughout an observing night.

Adaptive optics, briefly introduced in Chapter 14's discussion of ground-based astronomy's continued competitiveness with space telescopes, works through a broadly similar principle applied to a much smaller, faster-responding deformable mirror placed later in the telescope's optical path: by continuously measuring the specific pattern of atmospheric turbulence directly above the telescope at any given moment (often using a bright natural star, or an artificial "guide star" created by shining a separate laser into the upper atmosphere to stimulate a controlled glowing point at a known, fixed location) and adjusting the deformable mirror's shape to precisely counteract that measured turbulence many hundreds of times per second, these systems can dramatically sharpen the final image compared to what the same telescope would produce without correction, substantially closing the sharpness gap between ground-based and space-based observation that motivated much of the original case for building telescopes in space in the first place.

Both of these technologies — segmented mirror construction and adaptive optics correction — represent decades of sustained, expensive engineering development, and both depend, in different ways, on the underlying site conditions discussed earlier in this chapter: segmented mirror systems benefit from stable, low-turbulence sites because less correction is required to begin with, and adaptive optics systems, however sophisticated, work best, and in some configurations only work adequately at all, at sites with generally favorable atmospheric conditions to begin with, reinforcing rather than eliminating the fundamental importance of careful site selection this chapter has described throughout.

What happens when a site's conditions degrade after construction

The economic and scientific stakes discussed throughout this chapter become especially concrete when a site's conditions measurably degrade after a major facility has already been constructed and is already operating, since relocating an existing multi-billion-dollar observatory is, for all practical purposes, not a realistic option once construction is complete; the facility must either continue operating under degraded conditions, accepting some reduction in its achievable scientific performance relative to original expectations, or additional resources must be invested in mitigation measures — improved data processing to compensate for increased sky brightness or contamination, or, in genuinely severe cases, entirely new instrumentation designed specifically to work around degraded local conditions in ways the original facility design did not anticipate.

This is precisely the scenario the Paranal-region development dispute described in the previous chapter threatens to create in a particularly high-profile and consequential form: an existing, extraordinarily expensive, internationally significant observatory facility, built and sited specifically because of the surrounding region's exceptional darkness and atmospheric stability, facing the prospect of measurable degradation in exactly those founding conditions due to nearby development undertaken for entirely unrelated economic reasons. It is precisely this scenario — enormous sunk investment, genuinely difficult or impossible relocation, and a real risk of gradual, uncoordinated degradation of the specific conditions that justified the original investment — that gives the site-protection concerns discussed throughout this and the previous chapter their practical urgency, and that motivates the kind of dedicated, sustained national and international regulatory attention described in Chapter 9's discussion of Chile's observatory-protection framework.

The next chapter shifts from these ground-based stakes to an entirely different, and in some ways less familiar, category of threat to exactly the same underlying resource: darkness itself, arriving this time not from nearby terrestrial development but from thousands of kilometers overhead.

International collaboration as a funding and governance model

The international-consortium funding structure mentioned earlier in this chapter deserves slightly more elaboration, since it shapes not only how major observatories are financed but also how decisions about their operation, instrument allocation, and long-term strategic priorities are made. Major facilities of the scale discussed in this chapter are typically governed by formal international agreements among the participating national funding agencies and research institutions, establishing shared decision-making bodies responsible for allocating observing time among researchers from the various partner countries, setting long-term strategic and instrumentation priorities, and managing the facility's ongoing operational budget, which for a major observatory can itself run into tens of millions of dollars annually even after the initial capital construction cost has been fully paid.

This international governance structure has an important, if somewhat indirect, bearing on the site-protection concerns discussed throughout this chapter and the previous one: because a major observatory's governing partnership typically spans multiple national governments, any effort to secure additional environmental or lighting protections for the surrounding region — of the kind Chile has pursued at a national regulatory level, as discussed in the previous chapter — often benefits from this same multinational institutional backing, giving the diplomatic and advocacy effort behind such protections a weight and credibility that a purely domestic or purely scientific advocacy campaign might struggle to match on its own. This institutional weight has been evident in the international attention and formal objections generated by the Paranal development dispute discussed in the previous chapter, where the response came not merely from Chilean scientific institutions alone but from the full multinational partnership of governments and research organizations with an operational and financial stake in the facility's continued scientific productivity.

What replacing a degraded site would actually cost

It is worth making explicit, in closing this chapter, just how asymmetric the cost calculation is between protecting an existing dark-sky observatory site and attempting to replace one whose conditions have significantly degraded. Protective measures — the lighting ordinances, buffer-zone regulations, and coordinated advocacy discussed throughout this chapter and the previous one — typically cost a small fraction of a major observatory's original construction budget to implement and sustain over the facility's operational lifetime. Replacing a facility whose site conditions have become unacceptably degraded, by contrast, would require essentially repeating the entire multi-billion-dollar site-selection, construction, and commissioning process described throughout this chapter, at a new location, with no guarantee that a comparably excellent and comparably securely protected site remains available to select from in the first place, given the finite and already largely identified and claimed inventory of globally exceptional astronomical sites this chapter has described. This asymmetry — protection is comparatively cheap; replacement is enormously expensive and may not even be genuinely possible — is, in miniature, the same basic economic logic that underlies environmental protection arguments across many other domains, and it applies with particular force to the specific, finite, and irreplaceable resource this chapter has been concerned with throughout.

Training the next generation of observatory scientists

A further, less frequently discussed dimension of major observatory operation involves the specialized workforce required to design, build, and operate these facilities over their multi-decade operational lifespans: optical and mechanical engineers, software developers building the data pipelines discussed in Chapter 12, and the research astronomers themselves, whose training typically involves years of specialized graduate education. The scale and complexity of the facilities discussed in this chapter has driven, over recent decades, a corresponding expansion in specialized university programs and international training partnerships intended to build and sustain this workforce, an institutional investment that itself represents a further, less visible but genuinely significant component of the overall economic and human capital commitment underlying the observatory infrastructure this chapter has described.

Weather forecasting and observing-time efficiency

A further, practically significant aspect of major observatory operation involves the sophisticated weather and atmospheric-condition forecasting these facilities depend on to schedule their most demanding observations for the nights most likely to offer optimal seeing and clarity, discussed briefly earlier in this chapter. Given the enormous cost of observing time on a major facility of the scale this chapter has described, even modest improvements in forecasting accuracy can meaningfully improve a facility's overall scientific output by better matching its most sensitive, most weather-dependent observing programs to the specific nights actually best suited to them, an operational efficiency consideration that compounds, rather than substitutes for, the fundamental site-quality considerations this chapter has otherwise emphasized.

Remote operation and reduced on-site footprint

Modern major observatories increasingly rely on remote and queue-scheduled observation, in which researchers submit observing requests in advance and receive their data without physically traveling to the remote site at all, a shift that has, as a secondary benefit, reduced the on-site staffing and associated support infrastructure - and therefore the incidental lighting footprint - required at many major facilities compared to earlier eras when resident astronomers and larger visiting teams required more extensive on-site accommodation and support lighting.

Chapter 11 — The Sky Fills Up

Every threat to the night sky discussed so far in this book has come from the ground: cities, highways, buildings, signage, all of it fixed in place, all of it subject, at least in principle, to the kind of local policy interventions discussed in Chapter 9 and returned to at greater length in Chapter 15. A resident of a heavily light-polluted city has always retained one theoretical option, however impractical it may have become in practice as skyglow domes expanded and merged: travel far enough, and eventually you reach genuine darkness again. This chapter introduces a threat that breaks that assumption entirely, because it does not stay on the ground. It orbits.

The rise of the satellite megaconstellation

Artificial satellites have orbited Earth since 1957, and for the first six decades of the space age, the total number of operational satellites in orbit at any given time remained, by the standards of what would soon follow, quite modest — numbering in the low thousands at most, spread across a range of orbital altitudes and serving a correspondingly wide range of purposes: communications, weather monitoring, scientific research, navigation, and military applications among them. Visible artificial satellites were, for most of this period, a relatively occasional feature of the night sky: a careful observer might spot several in the course of an evening's stargazing, generally appearing as a single steadily moving point of light crossing the sky over a period of a few minutes, an interesting but not disruptive addition to an otherwise natural view.

That situation changed dramatically and very rapidly beginning in 2019, when a single company began launching satellites intended to provide global broadband internet access from a low Earth orbit constellation numbering, in its initial approved configuration, several thousand individual satellites — and in subsequent regulatory filings, potentially tens of thousands more. The scale of this single company's launch program alone exceeded, within a few years, the total number of satellites launched by all previous space-faring nations and companies combined across the entire six-decade history of spaceflight up to that point. By the mid-2020s, several thousand of this company's satellites were operational in orbit, with continued regular launches adding more, and at least two additional companies had begun deploying, or had received regulatory approval to deploy, competing broadband satellite constellations of their own, at a broadly comparable scale.

Why so many, and why so bright

The underlying business logic driving this rapid, large-scale deployment is straightforward and, on its own terms, compelling: providing broadband internet access to the substantial fraction of the world's population currently lacking reliable terrestrial broadband infrastructure — rural and remote regions poorly served by existing fiber and cellular networks — represents both a genuine unmet need and a very large potential commercial market. Low Earth orbit satellite constellations offer a technically appealing solution because satellites at these relatively low altitudes (typically a few hundred kilometers above Earth's surface, dramatically lower than traditional geostationary communications satellites) can provide comparatively low-latency internet connectivity, competitive with terrestrial broadband in ways that older, higher-altitude satellite internet systems could never quite achieve. Achieving continuous global or near-global coverage at these lower altitudes, however, requires a very large number of individual satellites, since each satellite at low altitude covers a correspondingly smaller area of Earth's surface at any given moment and orbits comparatively quickly, meaning constant handoffs between overhead satellites are required to maintain continuous coverage for any given ground location — which is precisely why the constellations involved number in the thousands rather than the dozens.

The visibility of these satellites to naked-eye and telescopic observers depends on their size, orbital altitude, and, critically, their surface reflectivity — how much sunlight they reflect back toward Earth-based observers, which depends on their physical design, the materials used in their construction, and their orientation relative to the sun and the observer at any given moment. Newly launched satellites, before reaching their final operational orbit and orientation, have in many cases been reported as unusually bright, appearing to observers as a distinctive "train" of evenly spaced points of light moving across the sky together shortly after a launch — a phenomenon that generated considerable public attention and, in the astronomical community, significant concern when the practice first became widely visible beginning in 2019, since a single launch could place dozens of new, close-together, comparably bright objects into the sky at once. Even after satellites reach their final operational configuration and are typically somewhat dimmer than during this initial post-launch period, they generally remain conspicuously visible to unaided observers under the right conditions — typically shortly after sunset or before sunrise, when the satellite remains in sunlight while the observer's local sky has already darkened — and the sheer number of them means that a given patch of sky, at a given time of night, now regularly contains multiple visibly moving satellites where a comparable observation a decade earlier might have shown one or none.

The regulatory vacuum

Perhaps the most striking feature of this rapid transformation, from the perspective of a book primarily concerned with policy and institutional response, is how little existing regulatory infrastructure was actually positioned to address it before it began. Satellite launches and orbital operations are subject to a considerable body of existing international and national regulation, covering matters including radio spectrum allocation (satellites communicate using specific radio frequencies that must be coordinated to avoid interference with other users), orbital debris mitigation guidelines (addressing the risk of collisions and the accumulation of space debris, a related but distinct concern from the visibility issue this chapter focuses on), and national licensing requirements administered, in the United States, primarily through the Federal Communications Commission for the communications aspects of these systems.

Strikingly, however, essentially none of this existing regulatory framework was designed with visual or astronomical impact specifically in mind, because the scale of satellite deployment that would eventually create a meaningful visual and astronomical impact simply did not exist, and was not seriously anticipated, at the time most of this regulatory structure was established. Regulatory review processes for new satellite constellation licenses have, since 2019, begun incorporating at least some consideration of astronomical impact, generally as a result of direct advocacy and formal comment submissions from professional astronomical organizations rather than as a pre-existing formal requirement built into the licensing process from the start, and the resulting mitigation commitments — discussed further in Chapter 15 — have generally been voluntary undertakings by the satellite operators themselves rather than binding regulatory requirements imposed by any government agency with clear jurisdictional authority over the specific question of night-sky brightness.

This regulatory gap reflects a deeper and genuinely difficult underlying legal reality: no single national government or international body holds clear, established jurisdiction over the specific question of how much artificial brightness or visual clutter is an acceptable addition to the shared night sky as experienced from Earth's surface, in the way that, for example, national environmental regulators hold reasonably clear authority over air and water pollution within their borders. The night sky, unlike the air above a specific nation's territory or the water within its borders, is experienced globally and continuously by observers in every country simultaneously, while the satellites causing the disputed effect may be licensed and operated by a company based in a single country, launched from another, and visible from literally every populated location on the planet at one time or another as their orbits carry them around the globe. No existing international legal framework was built to adjudicate a dispute with quite this structure, and the result, as of this book's writing, is a situation in which a rapidly escalating change to a genuinely global shared resource is proceeding largely on the basis of individual company launch schedules and voluntary mitigation commitments, rather than any coordinated international regulatory framework specifically designed to weigh the competing interests involved.

The astronomical community's response

The professional astronomical community's response to this rapid deployment has been active and, at times, urgent in tone, reflecting genuine alarm within the field about the potential scientific consequences, discussed in detail in the next chapter. The International Astronomical Union, the primary global professional body representing professional astronomers, established a dedicated center specifically to coordinate research and advocacy on the satellite constellation issue, reflecting a judgment within the field that the problem merited a level of institutional response comparable to other major, field-wide scientific concerns. Individual observatories and major upcoming astronomical facilities, particularly wide-field survey telescopes designed to repeatedly image large areas of sky (whose vulnerability to satellite interference is discussed at length in the next chapter), have engaged directly and extensively with satellite operators, providing detailed technical data on satellite brightness and trajectory to inform both operator-side mitigation efforts and the broader public and regulatory debate over the issue.

This engagement has, to be clear, produced some concrete results, discussed further in Chapters 13 and 15: certain satellite operators have adopted brightness-reducing design changes to at least some of their satellites, and coordination mechanisms allowing astronomers to receive advance notice of satellite positions, so that observations can be timed or processed to minimize interference, have been developed and refined. Whether these mitigation efforts are sufficient to offset the sheer scale of continued deployment, however — with total planned satellite numbers across all currently proposed constellations potentially reaching well into the tens of thousands over the coming decade — remains a genuinely open and actively debated question within the astronomical community, and is the specific subject of Chapter 13's discussion of where current trends might lead.

The next chapter turns from this broad historical and regulatory overview to a more technical and specific question: what, precisely, do these satellites actually do to a telescope's data, and why does even a seemingly modest number of bright, fast-moving objects crossing the field of view of a sensitive astronomical instrument create a disproportionately large scientific problem.

A brief technical primer on low Earth orbit

To understand why the megaconstellation deployment described in this chapter takes the specific form it does, it helps to understand a few basic facts about the orbital regime these satellites occupy. Low Earth orbit, generally defined as the region extending from roughly 160 to 2,000 kilometers above Earth's surface, is dramatically closer to the ground than the geostationary orbit historically used by many traditional communications satellites, which sits at approximately 35,786 kilometers altitude, a distance chosen specifically because a satellite at that precise altitude orbits at the same rate the Earth rotates, allowing it to remain fixed over a single point on the ground indefinitely — extremely useful for certain fixed communications applications, but involving a considerably longer round-trip signal delay (latency) than a much closer low-Earth-orbit satellite, given the simple physical fact that radio signals, like light, travel at a fixed speed and take measurably longer to cover a much greater distance.

This latency difference is the core technical reason the current generation of broadband-focused satellite constellations operates in low orbit rather than the traditional geostationary band: a low-orbit satellite can provide round-trip latency competitive with, and in some cases better than, traditional terrestrial broadband infrastructure, a capability existing geostationary satellite internet services, hampered by the much greater signal travel distance, historically struggled to match, limiting their usefulness for latency-sensitive applications like video calls or online gaming. The tradeoff for this latency advantage, as the main chapter text has already described, is that any single low-orbit satellite covers a much smaller area of Earth's surface at any given moment, and moves across the sky, relative to any fixed ground observer, very rapidly — completing a full orbit of the Earth roughly every ninety minutes, compared to the geostationary satellite's fixed, unmoving position — necessitating the very large total satellite counts this chapter has described in order to maintain continuous, handed-off coverage for users on the ground as individual satellites rise and set relative to their location throughout the day.

The "satellite train" phenomenon, explained

The especially striking, viral-video-friendly phenomenon of a bright, evenly spaced "train" of satellites crossing the sky together, mentioned briefly in the main chapter text, has a specific technical explanation worth spelling out. When a batch of satellites is launched together aboard a single rocket, they are typically released into a broadly similar initial orbit in close proximity to one another, and it takes a period of days to weeks, using each satellite's own onboard propulsion system, for them to gradually maneuver into their final, more widely distributed operational positions within the overall constellation. During this initial post-launch period, before this final spreading-out maneuver is complete, the newly launched satellites remain visible from the ground as a distinctive, evenly spaced line or "train," all moving together across the sky in formation, a visual effect striking and unusual enough that it has generated considerable public attention, viral social media video, and, on more than one occasion documented in press coverage, alarmed reports from members of the public unfamiliar with the phenomenon who mistook the formation for something else entirely.

This train effect is inherently temporary for any given launch batch, typically lasting from a few days to a few weeks before the satellites disperse into their final operational configuration and become correspondingly less visually striking, though the practical consequence of frequent, ongoing launches is that some degree of visible train formation has become a comparatively routine, recurring feature of the night sky in the years since large-scale megaconstellation deployment began, simply because new launches have continued occurring frequently enough that a newly launched, not-yet-dispersed train is very often visible somewhere in the sky on any given clear night, even after any specific individual train has long since dispersed.

Competing constellations and the broader commercial landscape

While this chapter and the ones following it necessarily focus most extensively on the single largest and earliest-deployed satellite broadband constellation, it is important to note that the broader commercial and competitive landscape now includes multiple substantial competing projects, each contributing independently to the cumulative satellite population and cumulative astronomical impact discussed throughout this book. At least two other major commercial operators have received significant regulatory authorization and begun active deployment of their own large constellations, pursuing broadly similar broadband connectivity business models, and several national governments, motivated by a combination of domestic connectivity goals and, in some cases, explicit strategic and sovereignty considerations around not depending entirely on a single foreign-operated satellite network for critical connectivity infrastructure, have announced or begun deploying their own state-backed or state-supported constellation projects.

This competitive proliferation matters directly for the cumulative-impact projections discussed in Chapter 13, since the astronomical community's concern is not limited to any single operator's satellites but extends to the total combined population across all currently operating and proposed constellations, a total considerably larger, and growing through a considerably larger number of independent commercial and national decision-making processes, than would be the case if a single operator's plans were the only relevant consideration. It also complicates the regulatory and diplomatic picture described later in this chapter and revisited in Chapter 13, since any eventual international coordination mechanism adequate to address the cumulative astronomical impact of satellite megaconstellations would need to account for multiple independent operators across multiple national jurisdictions, each with its own regulatory relationship to its home government and its own commercial incentives, rather than a single company whose behavior a single national regulator might address unilaterally.

Public perception and the "streaks in my photos" moment

Public awareness of the satellite megaconstellation issue, distinct from the more specialized professional astronomical concern discussed throughout this chapter, arguably crystallized for a broad general audience through a specific and widely shared type of experience: amateur photographers and astrophotographers, sharing images on social media of satellite trails unexpectedly crossing what they had intended as clean shots of stars, meteor showers, or other celestial subjects, generating viral attention and, in a number of documented cases, direct media coverage examining the phenomenon for a general audience unfamiliar with the underlying technical and regulatory background this chapter has provided. This grassroots, photography-driven public awareness has, in turn, fed back into the broader public and policy conversation about satellite constellation growth, providing a wider base of general public concern beyond the professional astronomical community's more technical, survey-telescope-focused advocacy discussed at greater length in the following chapter.

This dynamic is worth noting because it illustrates a recurring pattern in how environmental and technological harms enter public consciousness more broadly: often not through top-down expert communication alone, however important that expert communication ultimately is for shaping policy, but through a large number of individual, personally experienced encounters — an amateur photographer's ruined exposure, a family noticing an unusual bright light "train" during an evening walk — that collectively build a broader public intuition that something has changed, well before most of those individuals have engaged with the more technical scientific literature this book has drawn on throughout its second half.

Orbital altitude choices and their visibility tradeoffs

A further technical detail worth adding to this chapter's discussion concerns the specific choice of orbital altitude made by different constellation operators, which involves a tradeoff directly relevant to the visibility concerns this book addresses. Lower orbital altitudes generally produce somewhat dimmer satellites as seen from the ground (since the satellite's physical size subtends a smaller angle and its distance attenuates its reflected brightness more), and lower-altitude satellites also naturally decay out of orbit more quickly if they fail or reach the end of their operational life, reducing long-term orbital debris accumulation risk — but lower altitudes also mean each satellite covers a smaller ground area, requiring more satellites for equivalent coverage, and can result in a satellite spending less time in direct sunlight relative to the ground observer's darkness window, which can either increase or decrease the practical frequency of visible passes depending on the specific geometry involved. Different operators have made different altitude choices reflecting different weightings of these tradeoffs, and at least some public engagement between satellite operators and the astronomical community, discussed further in the next two chapters, has specifically addressed altitude choice as one lever, among several, potentially available for reducing the visibility and interference impact of future constellation deployment.

Public-private information sharing agreements

Beyond the coordination systems already discussed, it is worth noting that at least some satellite operators have entered into formal, if still relatively limited, information-sharing agreements with astronomical organizations, providing orbital ephemeris data (precise predicted position information) in advance of launch and throughout a satellite's operational lifetime, allowing observatories to model and, where practical, schedule around predictable interference. These agreements remain bilateral and voluntary rather than industry-wide binding requirements, and their scope and reliability have varied considerably between different operators, but they represent a genuine, if partial, institutional response to the concerns this chapter has documented, and a foundation that more comprehensive future regulatory frameworks, discussed in Chapter 13, could potentially build upon and formalize.

Ground-station infrastructure and its own lighting footprint

Beyond the orbital satellites themselves, it is worth briefly noting that the ground-station infrastructure required to communicate with large satellite constellations — sizable antenna installations typically requiring their own security and operational lighting — represents an additional, if comparatively minor, terrestrial lighting footprint connected to the broader megaconstellation phenomenon this chapter describes, illustrating that the orbital and terrestrial halves of this book's overall argument are not always as cleanly separable in practice as the book's two-part structure might otherwise suggest, since a fully orbital technology like satellite broadband still depends on a network of physical, light-emitting ground infrastructure to actually function.

Consumer terminal proliferation

The rapid growth in the number of consumer ground terminals used to receive satellite broadband service, numbering in the millions across the constellations discussed in this chapter, represents the demand-side counterpart to the orbital deployment this chapter has otherwise emphasized, and its continued growth is itself one of the commercial drivers sustaining operators' publicly stated confidence in continuing to expand their orbital fleets at the scale and pace this chapter has described.

Chapter 12 — A Telescope's Worst Enemy

It would be easy to assume, from a casual glance at the night sky, that a handful of satellites crossing among thousands of stars represents a minor, easily filtered annoyance rather than a serious scientific problem. This chapter explains, in some technical detail, why that assumption is largely wrong for a specific and increasingly important category of astronomical instrument: the wide-field survey telescope, and why satellite interference poses a disproportionately serious threat to precisely the kind of astronomy the field has been investing in most heavily in recent years.

How a modern survey telescope actually works

Traditional astronomical observation, for most of the twentieth century, typically involved pointing a telescope at a single object of interest for an extended exposure, gathering as much light from that one target as possible. A newer and increasingly important observational strategy, made possible by advances in digital detector technology and data processing capacity, instead involves systematically imaging very large areas of the sky repeatedly over time, building up a comprehensive, repeatedly updated map of everything visible in that area, rather than focusing on one predetermined target. This survey approach has become central to a wide range of contemporary astronomical priorities: detecting potentially hazardous near-Earth asteroids before they are discovered by chance, identifying transient astronomical events (supernovae, gravitational-wave counterparts, and other short-lived phenomena) as close as possible to when they actually occur rather than after the fact, and building enormous statistical catalogs of galaxies and stars precise enough to support large-scale cosmological research.

The most ambitious current example of this approach is the Vera C. Rubin Observatory, a large wide-field survey telescope constructed in Chile and designed to repeatedly image the entire visible southern sky every few nights over a planned decade-long survey program, generating an enormous, continuously updated astronomical dataset intended to support a broad range of scientific goals simultaneously, including precisely the kind of near-Earth-object detection, transient-event discovery, and large-scale cosmological mapping described above. Because this instrument is specifically designed to image very large areas of sky, repeatedly, with a wide-field camera capturing an enormous field of view in each individual exposure, it is disproportionately exposed to satellite interference compared to a traditional telescope observing a single small target: where a narrow-field telescope pointed at one specific distant galaxy might go an entire night without a satellite happening to cross its narrow field of view, a wide-field survey instrument imaging a large swath of sky on every single exposure has a correspondingly much higher probability that some satellite, somewhere in its enormous field of view, will cross during any given exposure.

What a satellite trail actually does to an image

When a satellite crosses a telescope's field of view during an exposure, it does not merely appear as a small dot the way a star does; because the satellite is physically close (a few hundred kilometers, compared to the effectively infinite distance of a star) and moving rapidly relative to the fixed background stars, it appears in a long-exposure image as a bright streak crossing some portion of the frame, its brightness often significantly exceeding that of the faint astronomical objects the exposure was actually designed to detect. Depending on the satellite's brightness and the sensitivity settings of the specific exposure, this streak can range from a minor, easily flagged artifact affecting a small fraction of the image, to a genuinely severe problem: modeling and early operational studies of the Rubin Observatory's survey specifically, examining the expected and observed impact of the current and projected future satellite population, have found that a meaningful fraction of individual exposures are affected by at least one satellite trail, with certain twilight observing periods — scientifically valuable specifically because they offer the best opportunity to detect near-Earth asteroids and other objects best seen against a partially lit sky, precisely the periods when low-orbit satellites remain most likely to still be in sunlight while the ground below has already darkened — showing disproportionately high rates of satellite contamination compared to the deep overnight hours.

The practical consequences of a satellite trail crossing an exposure vary depending on exactly where the trail falls and what the exposure was intended to detect. In some cases, sophisticated image-processing software can identify and mask the affected pixels, effectively discarding the contaminated portion of the image while preserving the rest — a workable if imperfect mitigation that reduces, without eliminating, the effective usable area of a contaminated exposure. In other, more serious cases, particularly when a trail happens to cross precisely the location of a faint transient event (a newly appearing supernova, for example, or the optical counterpart to a gravitational-wave detection, both categories of event whose entire scientific value often depends on being detected and characterized as promptly as possible after they occur), the contamination can directly cause a genuine scientific detection to be missed, delayed, or degraded in a way that no downstream data processing can fully recover, since the information that would have been captured in that specific patch of sky during that specific exposure is permanently lost once the exposure is complete, unlike stellar light, which will simply be there again during the telescope's next scheduled pass over that same patch of sky.

Detector saturation and secondary artifacts

Beyond the direct streak itself, sufficiently bright satellite crossings can cause a related but distinct problem: detector saturation, in which the affected pixels receive so much light that the sensor's electronics are overwhelmed, producing not merely a bright line in the image but a range of secondary artifacts — bleeding of excess signal into neighboring pixels, temporary reduced sensitivity in the affected sensor region for a period after the saturating event, and, in some detector technologies, subtler systematic effects on the calibration of the entire image that can be more difficult to fully correct for than the visible streak itself. These secondary effects matter because they mean the practical impact of a bright satellite crossing is not always confined neatly to the pixels the satellite's trail actually occupies; depending on the specific detector technology and the satellite's brightness, the effective "damage" to an exposure's scientific usability can extend somewhat beyond the visibly obvious streak itself.

Why this specifically threatens time-domain astronomy

The category of astronomical research most acutely threatened by this pattern of interference is what researchers call time-domain astronomy: the study of astronomical objects and events that change over time, ranging from asteroids moving across the sky, to variable stars pulsing on regular cycles, to genuinely one-time transient events like supernovae and gravitational-wave counterparts that appear suddenly, evolve over a period of days to weeks, and then fade permanently. Because the entire scientific value of this category of research depends on capturing a specific, often rapidly evolving phenomenon during a limited observing window, the loss or degradation of even a single critical exposure — precisely the failure mode created by an inconveniently timed satellite trail — carries scientific costs that are structurally different from, and in some respects more severe than, the costs imposed by satellite interference on more traditional, single-target, extended-exposure observations, where a contaminated exposure can typically simply be discarded and repeated on a subsequent night without permanently losing the scientific opportunity.

Time-domain survey astronomy of exactly this kind, exemplified by the Rubin Observatory's planned decade-long survey program, has been identified by working astronomers and by formal published research — including recent peer-reviewed analyses specifically quantifying satellite trail rates in Rubin Observatory commissioning and early survey data — as one of the scientific priorities most disproportionately exposed to the growing satellite population, precisely because it depends on the combination of wide-field imaging (increasing the statistical likelihood that some satellite will cross any given exposure) and time-sensitivity (meaning a lost exposure frequently cannot simply be recaptured later without losing the scientific opportunity entirely) that satellite interference most directly undermines.

Mitigation efforts and their real-world limits

It would be inaccurate and unfair to suggest that nothing has been done to address this problem; a considerable amount of technical mitigation work has occurred since the scale of the issue became clear in the early 2020s, discussed in more detail in Chapter 15's broader solutions discussion. Software-based trail detection and masking has become considerably more sophisticated, reducing, though not eliminating, the practical impact of a given satellite crossing on the usability of the surrounding, uncontaminated portion of an exposure. Coordination systems allowing observatories to receive predicted satellite position data in advance, developed in direct collaboration between astronomical institutions and at least some satellite operators, allow some observing programs to schedule around predictable satellite crossings where the specific target and timing flexibility make that practical — though this approach obviously does not work for genuinely unpredictable transient events, which by definition cannot be scheduled around anything.

Physical satellite design changes — darkening coatings and sun-shade elements intended to reduce a satellite's reflectivity and therefore its brightness as seen from the ground, adopted by at least one major operator on at least some of its satellites following direct engagement with the astronomical community — have measurably reduced, though not eliminated, the brightness of the specific satellites incorporating these changes, and remain voluntary design choices rather than a binding requirement applied uniformly across every operator and every satellite currently being launched by any company.

Taken together, these mitigation efforts represent genuine, good-faith technical progress, but they operate against a backdrop, described in the previous chapter, of continued rapid deployment by multiple competing operators, several of which have made no comparable public mitigation commitments at all, and none of which currently operate under any binding regulatory requirement to do so. Whether the rate of technical mitigation improvement can keep pace with the rate of continued deployment growth — or whether, as the next chapter explores in detail, current trends point toward a satellite population large enough to overwhelm even substantially improved mitigation measures — is the specific, still-unresolved question this book turns to next.

A closer look at the Rubin Observatory's specific vulnerability

Because the Vera C. Rubin Observatory features so prominently in this chapter's discussion, it is worth describing its specific design and scientific mission in slightly more detail, since both directly explain why it has become the emblematic case study for satellite interference concerns within the professional astronomical community.

The observatory's defining instrument is an extremely large digital camera — by some measures the largest and highest-resolution camera ever built for astronomical purposes — paired with a wide-field telescope design specifically optimized to image very large areas of sky in each individual exposure, in service of a planned observing program intended to image the entire accessible sky repeatedly, every few nights, over a survey lifetime of roughly a decade. This combination of extremely wide field of view and extremely frequent, repeated imaging of the same sky areas is precisely what makes the facility so scientifically powerful for detecting transient and moving objects — asteroids, supernovae, and other time-variable phenomena that a less frequently repeated survey would be more likely to miss entirely — and precisely what makes it so disproportionately exposed to satellite trail contamination compared to more traditional, narrower-field instruments, for exactly the statistical reasons described earlier in this chapter: a wider field of view crossing a larger area of sky in every exposure has a correspondingly higher probability that some satellite, somewhere within that larger area, will be present during any given exposure.

Early commissioning-phase data from the observatory, and subsequent peer-reviewed analysis of that data specifically examining satellite trail rates, has provided some of the most rigorous, directly measured (rather than purely theoretically modeled) quantification available anywhere of exactly how significant the satellite interference problem has already become for this specific category of instrument, given the satellite population already in orbit at the time that data was collected — data referenced earlier in this chapter and returned to again in the projections discussed in the next chapter, since it provides a real, empirically grounded baseline against which future, larger-constellation scenarios can be more confidently modeled, rather than relying purely on theoretical projection with no directly observed baseline for comparison.

Why "just process it out" is not a complete solution

It is worth directly addressing a common, intuitively appealing but ultimately incomplete response to the satellite interference problem described throughout this chapter: the suggestion that sufficiently sophisticated image-processing software should simply be able to identify and remove satellite trails from affected images, restoring the underlying astronomical data as though the interference had never occurred.

This intuition is not entirely wrong — as this chapter has already noted, trail-detection and masking software has genuinely improved considerably since the scale of the problem became clear, and does meaningfully reduce the practical impact of a great many satellite crossings. But it faces a hard, physical limit that no amount of software sophistication can fully overcome: once a satellite trail has saturated or otherwise corrupted the specific pixels covering a specific patch of sky during a specific exposure, the actual astronomical information that would have been captured in that specific patch of sky, during that specific moment in time, is genuinely and permanently lost, not merely obscured behind a removable artifact. Masking software can identify the affected pixels and prevent them from being mistakenly interpreted as a stray astronomical signal, and it can, in some cases, allow the affected region to be filled in using data from a different, uncontaminated exposure taken at a different time — but this substitution works only for objects that do not change between the two exposures, which is precisely the opposite of the sort of unpredictable, evolving transient events that time-domain survey astronomy is most interested in capturing in the first place. For the specific type of science this chapter has emphasized as most acutely at risk — genuinely one-time, unpredictable transient events, observed during a specific, non-repeatable moment — no amount of downstream processing sophistication can recover data that a satellite trail happened to obscure during the one specific exposure that mattered.

This distinction — between interference that can be adequately compensated for after the fact, and interference that represents a genuine, permanent loss of otherwise-unrecoverable scientific information — is the single most important technical point this chapter has tried to establish, and it is the reason the astronomical community's concern about satellite interference cannot be fully resolved through software improvements alone, however valuable those improvements genuinely are as a partial mitigation. Addressing the underlying problem at its source — reducing the number and brightness of interfering satellites themselves, through the design and coordination measures discussed at the end of this chapter and revisited in Chapter 15 — remains necessary precisely because no downstream processing fix can substitute for it completely.

Twilight astronomy and the near-Earth-object detection mission

This chapter has already noted that twilight observing periods carry disproportionate scientific importance for near-Earth object detection while also suffering disproportionate satellite contamination, and the specific reasons twilight matters so much for this mission deserve slightly fuller explanation, since planetary defense — the detection and tracking of asteroids that could potentially impact Earth — is among the most publicly resonant scientific goals served by the wide-field survey telescopes this chapter has discussed at length.

Asteroids whose orbits bring them closer to the Sun than Earth's own orbit are, by definition, most observable from Earth's surface during the twilight hours shortly after sunset or before sunrise, when the sky background is dark enough for meaningful observation but the observing geometry still allows a view of the sky region closer to the Sun's direction where these specific objects are more likely to be positioned at any given time. This same twilight window, as this chapter has already noted, is precisely when low-orbit satellites remain most likely to still be illuminated by direct sunlight even as the ground below has fully darkened, creating an unfortunate coincidence between the observing conditions most valuable for a specific, high-stakes scientific and public-safety mission and the observing conditions most heavily contaminated by satellite interference. Researchers and planetary-defense advocates within the astronomical community have specifically flagged this twilight-window overlap as one of the more concerning specific consequences of continued unmitigated satellite population growth, given the direct public-safety relevance of the near-Earth-object detection mission this specific observing window serves.

Data pipeline costs: an economic dimension often overlooked

Beyond the direct scientific costs of lost or degraded observations discussed throughout this chapter, it is worth noting a related economic cost that receives less public attention: the increased computational and engineering cost of developing, maintaining, and continuously improving the satellite-trail detection and masking software this chapter has discussed as a partial mitigation. Major survey telescope projects of the scale discussed in this chapter maintain substantial dedicated data-processing pipelines and software engineering teams, and a meaningful and growing fraction of that computational and engineering investment is now specifically directed at satellite-interference mitigation — work that would simply not have been necessary, or would have been necessary to a much smaller degree, in the pre-megaconstellation era of ground-based astronomy. This represents a genuine, if somewhat diffuse and difficult-to-precisely-quantify, additional cost imposed on the broader astronomical research enterprise by satellite constellation growth, above and beyond the more direct costs of lost or degraded scientific data already discussed at length, and it is a cost ultimately borne by the same public funding agencies and international partnerships responsible for financing the observatories described in Chapter 10, rather than by the satellite operators whose deployment activity has necessitated the additional mitigation investment in the first place.

Spectroscopy and a further vulnerable observation type

Beyond the imaging-survey vulnerabilities already discussed at length in this chapter, it is worth briefly noting that spectroscopic observations — which spread a target's light across a range of wavelengths to analyze its chemical composition, velocity, and other physical properties, rather than simply recording its brightness and position — face a related but distinct vulnerability to satellite interference, since a satellite crossing a spectrograph's narrow slit or fiber-fed input during an exposure can contaminate the resulting spectrum in ways that are often more difficult to identify and correct for than a simple imaging-survey streak, given the more complex, wavelength-dependent nature of spectroscopic data. This represents a further, less publicly discussed but scientifically meaningful category of satellite-interference risk beyond the wide-field imaging-survey case that has dominated this chapter's discussion.

Independent replication and the credibility of interference findings

It is worth noting, in support of the credibility of the specific interference findings this chapter has relied on, that satellite trail contamination research has been conducted and independently replicated by multiple separate research groups using different telescopes, different data-processing pipelines, and, in some cases, different specific satellite populations under study, generally converging on broadly consistent conclusions about the scale and severity of the interference problem for wide-field survey astronomy. This kind of independent, cross-validated replication is exactly the standard scientific practice that gives a body of research findings genuine credibility, and its presence here should reassure readers that this chapter's central claims rest on considerably more than a single research group's isolated analysis.

Archival data and retrospective reanalysis

Because major survey telescopes retain their raw exposure data indefinitely, researchers have been able to retrospectively reanalyze older observations using improved satellite-trail detection algorithms developed after the original data was collected, in some cases recovering scientifically usable information from exposures originally flagged as too contaminated to use, a promising if only partial mitigation avenue that depends entirely on the underlying raw data having been preserved rather than discarded at the time of original contamination.

International collaboration on trail-detection standards

Astronomical software developers across multiple countries and institutions have increasingly collaborated on shared, open-source satellite-trail detection tools, avoiding costly duplicated development effort and allowing smaller observatories without the resources of a flagship facility like Rubin to benefit from mitigation techniques originally developed for the largest, best-funded survey projects, a collaborative software-development pattern that mirrors similar open-source cooperation elsewhere in modern observational astronomy.

Chapter 13 — One Million Satellites?

Every number cited so far in this book's discussion of satellite megaconstellations has been a snapshot of a rapidly moving target. This chapter attempts something more difficult and necessarily more speculative: extrapolating from currently disclosed plans, regulatory filings, and industry trends to ask what the sky might actually look like a decade or two from now, if current trajectories continue roughly as currently announced — while being appropriately careful about the difference between a regulatory filing, a stated corporate ambition, and an actual, physically deployed and operational satellite.

Reading regulatory filings against deployment reality

A useful and important distinction runs through any attempt to project future satellite numbers: the gap between total satellites a company or government has filed for regulatory authorization to launch, potentially over a period of many years, and the number actually operational in orbit at any given moment. Regulatory filings, particularly with the International Telecommunication Union (which coordinates international radio spectrum allocation, a resource satellite communications depend on) and national regulators like the U.S. Federal Communications Commission, are frequently made well in advance of any firm commitment to build and launch the full authorized number, both because the regulatory process itself can take years and because companies have a strategic incentive to secure spectrum and orbital authorization rights before a competitor does, even for plans that may ultimately be scaled back, delayed, or never fully realized.

With that caveat firmly in place, the scale of currently authorized and proposed plans across the industry as a whole is nonetheless striking. The satellite operator responsible for the initial and largest current constellation had, as of the mid-2020s, regulatory approval for tens of thousands of satellites in its currently authorized configuration, well beyond the several thousand already operational, and had filed for additional authorization covering an even larger eventual total. At least two other major operators have received approval for, and begun deploying, competing constellations of comparable scale, and additional national and commercial projects — including state-backed constellations from other countries pursuing similar broadband or strategic communications goals — have been announced or are in early deployment stages. Adding these figures together, credible published estimates from industry analysts and astronomical researchers examining the cumulative regulatory landscape have suggested that the sum of currently authorized and proposed satellite constellations, across all operators, could plausibly exceed one hundred thousand individual satellites over the coming one to two decades if a substantial fraction of currently filed plans are actually realized — a scenario in which the informal shorthand "a million satellites," while almost certainly an exaggeration of any currently disclosed specific plan, no longer sounds like pure hyperbole so much as a plausible extrapolation of currently visible industry momentum, particularly once additional operators not yet publicly disclosed are considered.

What a genuinely large constellation population would mean for visibility

Modeling work published by researchers specifically examining the visual and astronomical impact of large future satellite populations — including a notable 2025-2026 modeling study extending existing analysis to significantly larger hypothetical constellation sizes than the currently operational population — has attempted to project what naked-eye and telescopic conditions would actually look like under various future deployment scenarios. The general conclusion of this body of work, expressed with appropriate scientific caution about the uncertainty inherent in any such projection, is that a sufficiently large total satellite population could plausibly reach a point where, at any given moment during the twilight hours when satellites remain most visible, a naked-eye observer at a mid-latitude location might expect to see multiple moving satellites simultaneously visible in the sky at once, essentially continuously, rather than as an occasional, discrete event — a genuinely qualitative change in the ordinary experience of looking at the night sky during those hours, transforming satellite visibility from an interesting occasional occurrence into a routine, more or less constant feature of the twilight sky.

For professional astronomy, the implications modeled in this research are, if anything, more severe, precisely for the reasons detailed in the previous chapter regarding wide-field survey telescopes: a sufficiently large total satellite population, distributed across the range of orbital altitudes and inclinations current and proposed constellations actually occupy, could push the fraction of wide-field survey exposures affected by at least one satellite trail considerably higher than the already meaningful rates observed with the current, still-partial deployment of currently authorized constellations, with some modeled high-deployment scenarios suggesting that certain especially valuable observing windows — again, particularly the twilight periods most useful for near-Earth object detection — could become so heavily contaminated that specific categories of scientific observation become substantially less practical to conduct from the ground at all during those windows.

Kessler syndrome: a related but distinct danger

Discussion of very large satellite populations inevitably raises a second, related concern that, while distinct from the visibility and light-pollution focus of this book, deserves at least a clear explanation given how frequently it appears alongside megaconstellation coverage: the risk of orbital debris cascades, sometimes referred to by the informal shorthand "Kessler syndrome," after the NASA scientist who first formally described the underlying mechanism in the 1970s. The basic concern is that as the total number of objects in a given orbital region increases, so does the statistical probability of collisions between them, and each collision — whether between two operational satellites, a satellite and existing debris, or two pieces of debris — generates additional fragments, each of which becomes a further collision risk to everything else sharing that orbital region, in principle creating a self-reinforcing cascade that could, in a sufficiently severe scenario, render certain orbital altitudes hazardous enough to discourage or prevent future satellite operations there for an extended period.

This is, to be clear, a different specific concern from the visual and astronomical brightness issues that are this book's primary focus, and it operates on a different technical basis, though both concerns are driven by the same underlying trend — a rapidly increasing total population of objects in low Earth orbit — and both have prompted overlapping calls from within the space and astronomical communities for more robust orbital traffic management and debris mitigation requirements than currently exist in binding international law. Readers interested in the debris and collision-risk dimension specifically should understand it as a related but analytically separate set of concerns from the light-pollution and astronomical-interference focus of this book, even though the two issues frequently appear together in both scientific literature and general media coverage of the megaconstellation phenomenon, and even though both point toward a common underlying policy gap: the absence of binding international coordination mechanisms adequate to the actual scale of contemporary satellite deployment.

The international law gap, revisited

Chapter 11 introduced the basic problem that no clear jurisdictional authority currently exists to weigh, on behalf of the global public that shares a single night sky, the astronomical and visual costs of satellite megaconstellations against their considerable communications and connectivity benefits. This chapter's extrapolation into plausible future deployment scenarios makes the practical stakes of that gap considerably more concrete: if current trends continue roughly as currently filed and announced, the decisions that will determine what the night sky looks like for the next several generations of human beings are being made, in practice, primarily by a small number of private companies and a handful of national telecommunications regulators, evaluating primarily commercial and spectrum-allocation criteria, with astronomical and general public interest in preserving a visually undisturbed night sky entering the process, where it enters at all, mainly through voluntary consultation and public comment processes rather than through any binding legal requirement to weigh that interest against the commercial and connectivity case for continued deployment.

This is not, to be clear, an argument that satellite broadband constellations serve no legitimate purpose or should not exist at all; the case for extending reliable internet connectivity to underserved regions is genuine and significant, and this book has no interest in dismissing it. It is, rather, an observation that a policy decision with genuinely global, multi-generational consequences for a shared resource — the visual character of the night sky itself — is currently being resolved almost entirely through market and regulatory mechanisms designed for much narrower and more conventional purposes (spectrum allocation, national telecommunications licensing), without any international mechanism specifically designed to weigh the full range of interests actually at stake, in the way that, for example, international environmental treaties have attempted, with varying success, to coordinate global response to other shared, borderless environmental concerns.

Holding uncertainty honestly

It is worth closing this chapter by restating, deliberately, the appropriate level of confidence to attach to everything in it. The specific numerical projections discussed here — potential total constellation sizes reaching well into the tens of thousands or more, modeled future visibility and survey-contamination scenarios under various deployment assumptions — are genuinely uncertain, dependent on business decisions, technological developments, and regulatory outcomes that have not yet occurred and could unfold in a range of different directions from what current trends alone would suggest. Some currently filed regulatory plans will likely be scaled back, delayed, or abandoned entirely, exactly as has happened with a range of ambitious satellite constellation proposals in previous decades that never reached their originally filed scale. Continued and possibly accelerating technical mitigation efforts, of the kind discussed at the end of the previous chapter, could meaningfully offset at least some of the visibility and interference impact of whatever total satellite population is eventually deployed.

What is not genuinely uncertain, however, based on the deployment that has already occurred and is already contractually and technically committed, is the basic direction of travel: the total number of satellites in low Earth orbit is considerably larger today than at any previous point in the history of spaceflight, is very likely to continue growing substantially over at least the next several years given deployment that is already underway and contracted, and has already produced measurable, documented effects on both naked-eye sky appearance and professional astronomical data quality, as the previous two chapters have detailed. The open question this chapter has tried to explore honestly is not whether this trend will continue at all, but how far it will ultimately go, and what that trajectory implies for the very question this book's next chapter takes up directly: whether, and under what conditions, ground-based astronomy itself remains a viable long-term strategy for observing the universe from the surface of an increasingly crowded planet.

How researchers actually build these projections

It is worth briefly explaining the general methodology behind the kind of future-scenario modeling referenced in this chapter, since understanding how these projections are constructed helps calibrate appropriate confidence in their conclusions. Researchers modeling future satellite visibility and interference impact typically begin from a set of assumptions about total future constellation size, drawn from some combination of currently authorized regulatory filings, publicly stated corporate deployment targets, and, for the most speculative "high-end" scenarios, extrapolation of current industry growth trends forward in time under an assumption of continued, unconstrained expansion.

From this starting population assumption, researchers apply known or estimated satellite brightness distributions (drawing on actual observed brightness measurements of currently operational satellites, adjusted for any known or planned design changes such as the darkening coatings discussed in the previous chapter), combined with orbital mechanics calculations determining how many satellites would be above the horizon and illuminated by sunlight at any given time from a given ground location, to estimate the resulting naked-eye and telescopic visibility conditions under each modeled scenario. This is a well-established and methodologically reasonably sound approach to projection under uncertainty, broadly comparable in its basic logic to the scenario-modeling techniques used in fields ranging from climate science to epidemiology, where firm future data does not yet exist and researchers must instead construct a range of plausible scenarios bounded by current trends and stated assumptions, rather than a single definitive prediction.

The appropriate way to read the specific numerical outputs of this kind of modeling, as this chapter has already emphasized, is not as precise forecasts of what will definitely occur, but as a rigorous exploration of what current trends, if they continued largely unchanged, would plausibly imply — valuable specifically because it makes the stakes of continued unmitigated deployment considerably more concrete and quantifiable than a purely qualitative warning could achieve, while remaining explicitly and appropriately conditional on assumptions that could, in practice, turn out differently than currently projected.

The view from an amateur astronomer's backyard

Beyond the professional, survey-telescope-focused concerns that have dominated this and the previous chapter, it is worth briefly returning to the experience of amateur astronomers and ordinary stargazers, whose relationship to the growing satellite population, while receiving less institutional and policy attention than the professional astronomical concerns discussed at length above, represents a genuinely widespread and rapidly growing source of frustration within amateur astronomy communities.

Long-exposure astrophotography, a popular and increasingly technically accessible hobby pursued by a very large number of amateur enthusiasts using increasingly affordable digital camera and telescope equipment, is subject to essentially the same satellite-trail contamination problem described in professional, survey-telescope terms in the previous chapter, simply at a smaller and more individually variable scale: an amateur astrophotographer attempting a multi-minute exposure of a faint deep-sky object has a meaningfully increased chance, compared to a decade ago, of having that exposure crossed by one or more visible satellite trails, a frustration widely and frequently discussed within online amateur astronomy communities and astrophotography forums, where the practical workaround — taking multiple shorter exposures and digitally combining them, discarding any individual frames contaminated by a satellite trail — has become an increasingly standard, if additionally burdensome, part of the amateur astrophotography workflow specifically in response to the growing satellite population.

This amateur-level impact matters for the broader argument of this book because it illustrates that the consequences of satellite megaconstellation growth are not confined to an elite, specialized professional research community operating billion-dollar facilities in remote deserts; they extend directly into the hobby and recreational experience of a very large number of ordinary stargazers and astrophotography enthusiasts worldwide, adding yet another concrete, widely and personally experienced dimension to a set of concerns that might otherwise be dismissed, inaccurately, as an issue affecting only a small, specialized professional community with no direct relevance to ordinary people's actual experience of the night sky.

Insurance, finance, and the debris-risk dimension in brief

Returning briefly to the Kessler-syndrome debris concern introduced earlier in this chapter, it is worth noting an economic dimension that has begun attracting attention from the commercial space insurance and finance sector specifically: as the total population of active satellites and tracked debris objects in low Earth orbit has grown, the statistical collision risk faced by any individual satellite operator has correspondingly increased, and satellite insurance underwriters have begun incorporating orbital congestion risk more explicitly into their risk models and premium calculations, creating a market-based signal, independent of any government regulatory action, that at least partially reflects the growing congestion this chapter has described. Whether this market-based signal will prove sufficient on its own to meaningfully constrain future deployment growth, absent the kind of binding international regulatory coordination this chapter has argued remains largely absent, is a genuinely open question that a number of space-policy researchers have examined, generally concluding that market mechanisms alone are unlikely to fully substitute for coordinated regulatory action, given the well-known tendency of insurance and market mechanisms to underprice diffuse, long-term, and difficult-to-attribute collective risks of exactly the kind orbital debris accumulation represents.

What would meaningful international coordination actually look like

It is worth closing this chapter with a brief, concrete sketch of what genuinely adequate international coordination on the issues this chapter and the preceding two have raised might actually involve, even though, as this chapter has repeatedly emphasized, no such coordination currently exists in binding form. Space-policy researchers and astronomical advocacy organizations have generally pointed toward some combination of the following elements as necessary components of an adequate response: binding, rather than voluntary, brightness and design standards applied consistently across all operators and jurisdictions, rather than the current patchwork of individual operator commitments; a formal mechanism for weighing astronomical and general public sky-visibility interests explicitly within the satellite licensing process, rather than through informal consultation alone; and some form of international body with actual authority to coordinate among the multiple national regulators and multiple satellite operators this chapter has described, comparable in institutional weight to existing international bodies coordinating other shared, borderless resources such as radio spectrum allocation itself.

None of these elements currently exists in the form just described, and this chapter does not predict with any confidence that they will be established, or established quickly, given the considerable diplomatic, commercial, and jurisdictional complexity involved in constructing genuinely binding international coordination of this kind from what is currently a largely voluntary and fragmented regulatory landscape. What can be said is that the astronomical community and a growing number of space-policy researchers have begun actively articulating what such a framework might look like, which is itself a necessary precondition for eventually building it, even if the political and diplomatic work of actually establishing it remains, as of this book's writing, substantially ahead rather than behind.

Other nations' emerging space policy responses

Beyond the United States-centered regulatory discussion that has dominated this chapter, it is worth noting that other spacefaring nations and blocs have begun developing their own regulatory approaches to satellite constellation licensing, with varying degrees of explicit attention to astronomical and environmental impact. The European Union and several individual European national space agencies have engaged in preliminary policy discussions specifically referencing astronomical dark-sky concerns, while other major spacefaring nations have thus far addressed constellation licensing primarily through existing frameworks focused on spectrum and orbital-debris management, with comparatively less explicit attention to the visibility and astronomical-interference concerns this book has emphasized, illustrating that the international regulatory patchwork described throughout this chapter varies not just in degree but in basic approach from one jurisdiction to the next.

Historical precedent: spectrum allocation as a partial model

International radio spectrum allocation, coordinated through the International Telecommunication Union and its associated treaty framework, offers at least a partial historical precedent for the kind of binding international coordination mechanism this chapter has argued remains largely absent for orbital light pollution specifically. Spectrum allocation, like night-sky darkness, involves a genuinely shared, borderless resource whose use by any single actor can meaningfully affect users elsewhere in the world, and the international community has, over many decades, built a reasonably durable, binding coordination framework to manage it, suggesting that comparable international coordination mechanisms for orbital brightness and debris management, while not yet developed to a similar degree, are not inherently impossible to construct, given sufficient sustained diplomatic attention and political will.

Public opinion polling on satellite constellations

A handful of public opinion surveys conducted in recent years have specifically asked general-population respondents about awareness of and attitudes toward satellite megaconstellations, generally finding relatively low baseline awareness of the specific astronomical concerns this book has documented, but considerably higher support for mitigation measures once respondents are briefed on the underlying issues, a pattern consistent with the broader observation, made earlier in this chapter, that public engagement on this issue remains at a comparatively early stage relative to the scale of the underlying trend.

The role of scientific journals in sustaining attention

Sustained coverage of satellite-interference research in leading peer-reviewed journals, referenced at several points throughout this chapter and the two preceding it, has played a meaningful role in keeping this issue visible within the broader scientific community, ensuring that new findings continue to reach policymakers and the public through credible, rigorously reviewed channels rather than fading from attention after the initial wave of public concern that followed the first large-scale constellation launches in 2019.

Chapter 14 — The Last Observatory

This chapter poses, without pretending to fully resolve, the single most difficult and consequential question this book has been building toward across its previous thirteen chapters: given the combined pressure of terrestrial light pollution, encroaching development even at remote sites, and a rapidly growing satellite population interfering with observations from every location on Earth regardless of remoteness, at what point — if any — does ground-based astronomy stop being the most sensible way to look at the universe?

Resisting the tempting, overstated version of the question

It would be easy, and this book has deliberately resisted the temptation throughout, to frame this question in its most dramatic possible form: a flat claim that ground-based astronomy is finished, that Earth's surface is no longer a viable platform for serious astronomical observation, that the future of the field lies entirely in space. That framing is not merely imprecise; it is not well supported by the current state of the science, and a knowledgeable reader would be right to push back hard against it.

Ground-based telescopes retain genuine, substantial, and in some respects still-growing advantages over space-based alternatives that are worth stating clearly before considering the countervailing pressures. Physical size is the most straightforward: a telescope's light-gathering power and image resolution both scale directly with the diameter of its primary mirror, and building, launching, and deploying an extremely large mirror in space is vastly more difficult and expensive than building the equivalent mirror on the ground, where gravity, transportation limits, and structural engineering constraints, while still very real, are dramatically less severe than the constraints imposed by rocket payload dimensions and the need for a space telescope's optical structure to survive launch forces and then unfold or otherwise achieve its operational configuration in the vacuum of space, entirely without the possibility of the kind of hands-on servicing and repair that a ground-based facility can always, in principle, receive. The largest ground-based telescopes currently operating or under construction have primary mirror diameters several times larger than the largest space telescope ever launched, a gap that is not likely to close dramatically any time soon given the physical and economic realities of launching enormous, precisely figured optical surfaces into orbit.

Adaptive optics technology, meanwhile, has substantially addressed what was historically ground-based astronomy's single biggest disadvantage relative to space telescopes: atmospheric blurring. By using deformable mirrors that adjust their shape many times per second, guided by real-time measurement of atmospheric turbulence (often using an artificial laser guide star as a reference point), modern ground-based telescopes can now achieve image sharpness in at least some observing conditions that rivals or in certain respects exceeds what a comparably sized space telescope could achieve, substantially closing a gap that, as recently as a few decades ago, was considered one of the most fundamental and permanent advantages of space-based observation.

Cost, too, favors ground-based facilities considerably, even accounting for the very large sums discussed in Chapter 10: launching, operating, and, when necessary, replacing or upgrading a space telescope involves costs and risks — launch failure, the impossibility of physical repair for many space telescope designs, the fixed operational lifespan imposed by fuel and component degradation in a harsh space environment — that a ground-based facility, however remote and however expensive to build and maintain, simply does not face in the same way. A ground-based telescope can, in principle, be upgraded, repaired, and operated for many decades; a space telescope's operational lifespan is generally fixed and finite from the moment of launch, with no realistic prospect of physical servicing beyond a small number of missions specifically designed to accommodate it.

Taking the countervailing pressure equally seriously

Having stated the case for ground-based astronomy's continued viability as clearly and fairly as possible, it would be equally dishonest to understate the genuine, mounting pressure the preceding chapters have documented. Sky brightness at even some of the most historically prized dark-sky observatory sites has measurably increased over recent decades, both from the gradual encroachment of nearby development described in Chapter 9's discussion of the Atacama, and, more universally and more rapidly, from the satellite interference described in Chapters 11 through 13, which — unlike terrestrial light pollution — cannot be escaped by any degree of geographic remoteness, since satellites in low Earth orbit pass over every location on the planet's surface in the ordinary course of their orbits, given enough time.

For certain specific categories of astronomical research, particularly the wide-field, time-domain survey astronomy discussed at length in Chapter 12, the combination of these pressures is disproportionately severe precisely because that research depends on exactly the observing conditions — wide fields of view, continuously repeated observation, sensitivity to faint and often unpredictable transient signals — most vulnerable to both increasing sky brightness and satellite trail contamination. It is entirely reasonable, and consistent with the actual published concerns of researchers working in this specific subfield, to worry that certain scientific goals within this category may become significantly harder, more expensive, or in some specific cases impractical to achieve from the ground as current trends continue, even while it remains true that ground-based astronomy as a whole, across its full range of scientific applications, retains substantial continued viability.

The right question, more precisely stated

This is why the framing this book has adopted from its introduction onward — not "will ground-based astronomy end" but "are we approaching a point where the night sky itself becomes a limiting resource for ground-based astronomy" — is the more defensible and, ultimately, the more useful way to state the question. A limiting resource is not necessarily a resource that has run out; it is a resource whose scarcity begins to actively constrain what would otherwise be achievable, requiring increasingly costly workarounds, increasingly careful allocation, or increasingly difficult tradeoffs between competing uses, well before it disappears entirely. Water, agricultural land, and radio spectrum are all limiting resources in exactly this sense in various contexts around the world today, without having become literally unavailable; their scarcity manifests as increased cost, increased competition, and increased engineering effort required to accomplish goals that would have been comparatively straightforward when the resource was more abundant.

Sky darkness, on the evidence assembled across this book's previous thirteen chapters, is increasingly displaying exactly this profile for at least some significant categories of ground-based astronomical research. It has not disappeared, and current evidence does not support a claim that it will disappear entirely in the near future even at the most heavily affected observatory sites. But it is becoming scarcer, more expensive to preserve, and more actively contested — by nearby industrial development in the case of sites like Paranal, and by an rapidly growing orbital satellite population in the case of the field as a whole — in ways that are already measurably affecting the design, planning, and expected scientific output of major current and future observatories, and that credible projections, appropriately caveated as discussed in the previous chapter, suggest could become considerably more severe over the coming decades if current trends continue largely unmitigated.

The billion-dollar question, stated plainly

The chapter's introduction posed a deliberately provocative version of this question, borrowed from the earlier discussion that inspired this book's overall framing: how dark does the sky have to become before a billion-dollar observatory is not worth putting on Earth? It is worth being honest that this book cannot answer that question with a specific threshold, because no such precise, general threshold actually exists in the relevant science or economics; the answer depends enormously on the specific scientific goals a given facility is designed to pursue, the specific mitigation technologies available at the time a given facility is planned and built, and the specific trajectory of satellite deployment and terrestrial development at the specific site under consideration, all factors that are themselves evolving rapidly and somewhat unpredictably, as the previous chapter's discussion of deployment uncertainty made clear.

What can be said with confidence is that this question — not merely as an abstract philosophical exercise, but as a genuine, practical planning consideration — is now, for the first time in the history of ground-based astronomy, one that observatory planners, funding agencies, and international scientific consortia are actively required to weigh when designing and siting the next generation of major astronomical facilities, in a way that was simply not necessary for the astronomers who selected and built the great observatories of the twentieth century. That, more than any specific numerical projection this book has offered, is the real measure of how much has changed: a resource that was, for the entire prior history of the field, simply assumed to be available in whatever quantity a sufficiently remote and well-chosen site could provide, must now be actively planned around, mitigated for, and in some cases explicitly traded off against other considerations, as a matter of routine practice rather than a hypothetical future concern.

The final chapter of this book's main argument turns to the question of what can actually be done about all of this — on the ground and in orbit alike — before the epilogue returns, one last time, to the beam of light atop a casino in the Nevada desert with which this book began.

Hybrid strategies: splitting the difference

Rather than framing the future of astronomical observation as a binary choice between ground-based and space-based platforms, the actual strategic direction being pursued by most major astronomical institutions and funding agencies is explicitly hybrid, deploying each platform for the specific scientific goals it is best suited to, while actively working to mitigate each platform's specific weaknesses — light pollution and satellite interference for ground-based facilities, size and serviceability constraints for space telescopes — rather than treating either platform as a complete, standalone solution to the full range of contemporary astronomical goals.

Space telescopes have historically been deployed, and continue to be planned, disproportionately for specific scientific goals where their unique advantages — freedom from atmospheric distortion entirely, access to wavelengths (certain infrared and ultraviolet bands, most prominently) that Earth's atmosphere absorbs before they can reach a ground-based instrument at all, and freedom from the terrestrial and orbital light pollution concerns this book has discussed at length — provide a decisive advantage that no amount of ground-based mitigation could fully replicate. Ground-based facilities, by contrast, remain the platform of choice for goals where sheer light-gathering power (favoring the much larger mirrors achievable on the ground, as discussed earlier in this chapter), the ability to be serviced, upgraded, or entirely reinstrumented over a multi-decade operational lifespan, and lower absolute cost per unit of scientific capability, provide the decisive advantage.

This division of labor is not a new or unique response to the specific pressures described in this book; it reflects, in large part, the same basic strategic logic that has governed the field's platform choices since the earliest space telescopes were launched in the late twentieth century, well before satellite megaconstellations existed as a serious consideration at all. What has changed, as this chapter has tried to establish carefully, is not the basic logic of this division of labor, but the specific balance of costs and benefits on the ground-based side of the ledger, which has shifted, for at least some scientific goals, somewhat further toward the space-based option than would have been the case a few decades ago, without shifting so far as to eliminate ground-based astronomy's continued, substantial advantages for the many scientific goals where those advantages remain decisive.

Voices from within the field

It is worth closing this chapter by noting, briefly, that the tone of concern within the working astronomical community about the issues discussed in this and the preceding chapters is genuinely serious, and not merely a rhetorical device adopted for the purposes of this book's argument. Formal statements and position papers from major international astronomical bodies, including the International Astronomical Union's dedicated center on the dark and quiet skies issue referenced in Chapter 11, have used language explicitly describing satellite megaconstellation growth as a significant and rapidly escalating threat to the future of ground-based astronomy, while consistently avoiding the kind of absolute, "astronomy is finished" framing this chapter has explicitly and deliberately declined to adopt.

This combination — genuine, institutionally expressed alarm, combined with careful avoidance of overstated, absolute claims about ground-based astronomy's viability — reflects, I think, exactly the right posture for a book like this one to adopt as well, and it is the posture this chapter has tried to model throughout: taking the documented risks fully seriously, without exaggerating them into a more dramatic but less accurate story than the actual evidence supports. The final chapter of this book's main argument, which follows, turns to what can be done to keep that balance tilted as favorably as possible toward continued, robust ground-based astronomical capability, even as the pressures this book has documented continue to mount.

Public funding priorities and the politics of choice

It is worth acknowledging directly, in this chapter's discussion of the ground-versus-space question, that the eventual balance between continued investment in ground-based facilities and continued investment in space telescopes is not merely a technical or scientific question to be resolved by engineering analysis alone, but is also, inescapably, a political and budgetary one, decided in practice through the funding priorities of a relatively small number of national science agencies and international partnerships, each facing their own competing demands across the full range of scientific fields they support, of which astronomy is only one among many.

This political dimension matters because it means the specific balance this chapter has discussed — substantial continued advantages for ground-based facilities in some respects, growing pressure from light pollution and satellite interference in others — will be resolved, in practice, not purely on the basis of the technical tradeoffs this chapter has laid out, but through the accumulated funding decisions of these agencies over the coming years and decades, decisions that will themselves be shaped by exactly the kind of public and institutional awareness this book has tried to contribute to: a clearer, more widely shared understanding of what is actually at stake, technically and scientifically, in the ongoing degradation of ground-based observing conditions this book has documented throughout its second half.

A note on public engagement and citizen advocacy

Beyond the formal institutional and funding-agency channels just discussed, it is worth noting that public awareness and citizen advocacy have played, and likely will continue to play, a meaningful role in shaping the policy responses discussed throughout this and the surrounding chapters. The kind of viral amateur-photography moments discussed in the previous chapter, combined with sustained media coverage of the Rubin Observatory's satellite-contamination findings and similar research, has measurably raised general public awareness of issues that, only a few years earlier, were known almost exclusively within the specialized professional astronomical community itself. This broadened awareness has, in turn, supported increased public comment submission during the regulatory review processes discussed in Chapter 11, providing at least some counterweight, however modest relative to the commercial interests involved, to the largely commercially driven regulatory process that has otherwise governed satellite constellation licensing to date. Whether this growing public engagement proves sufficient to meaningfully influence the policy outcomes discussed throughout this chapter and the next remains to be seen, but it represents a genuine and, compared to just a few years ago, considerably more developed channel through which the concerns this book has documented can enter the broader public and political conversation.

Next-generation space telescope economics

It is worth noting, in closing this chapter's ground-versus-space discussion, that the cost of major space telescopes has itself grown substantially with each successive generation, a trend that complicates any simple assumption that space-based observation offers an straightforwardly cheaper escape from the ground-based pressures this book has documented. Recent major space telescope projects have required development budgets comparable to or exceeding the largest ground-based facilities discussed in Chapter 10, once launch, testing, and mission-operations costs are fully accounted for, suggesting that the ground-versus-space tradeoff this chapter has examined is unlikely to be resolved simply by space telescopes becoming a uniformly cheaper alternative, and will instead likely continue to be decided, for the foreseeable future, on the mission-specific, capability-driven basis this chapter has described throughout.

Amateur and citizen contributions to professional science

It is worth noting, in closing this chapter's discussion of the future of ground-based astronomy, that the field has a long and continuing tradition of meaningful amateur contribution to professional scientific goals — variable-star monitoring, supernova and comet discovery, and, increasingly, coordinated citizen-science projects analyzing publicly released survey data. This tradition matters for the ground-versus-space question this chapter has examined because it represents a category of scientific contribution that depends specifically on widespread public access to reasonably dark skies, of the kind this book's opening chapters described as historically universal and increasingly rare, suggesting that the stakes of the terrestrial light pollution and satellite interference trends this book has documented extend beyond the professional research community's own facilities and into this broader, historically productive amateur-professional collaborative tradition as well.

Interferometry as a complementary ground-based technique

Beyond the single-large-mirror telescopes emphasized throughout this chapter, ground-based astronomical interferometry - combining light from multiple separate telescopes to achieve angular resolution equivalent to a single much larger instrument - represents a further technique through which ground-based facilities continue to expand their effective capability without requiring the impractically large single mirrors discussed earlier, offering yet another reason the ground-versus-space question this chapter has examined resists any simple, permanent resolution in either direction.

Chapter 15 — Can We Turn the Lights Down?

Every chapter of this book so far has documented a problem. This chapter, deliberately, is organized around solutions — not because the problem has been solved, or is close to being solved, but because, unlike a number of other major environmental challenges facing the contemporary world, light pollution turns out to be unusually, almost uniquely, tractable wherever there is sufficient institutional will to address it, and there are enough genuine, documented successes at multiple scales to justify cautious, evidence-based optimism rather than pure alarm.

The good news buried in the physics

Recall from Chapter 5 the basic physical mechanism behind terrestrial skyglow: artificial light directed upward, or reflected upward off the ground and other surfaces, scatters off atmospheric particles and returns to the sky as a diffuse glow. This physical mechanism carries an important, genuinely encouraging implication that distinguishes light pollution from many other categories of environmental harm: unlike, for example, carbon emissions, whose warming effect persists in the atmosphere for decades to centuries after emission, light pollution's effect is essentially immediate and immediately reversible at the source. A poorly shielded, upward-leaking streetlight replaced tonight with a properly shielded, fully downward-directed fixture stops contributing to skyglow the moment the replacement is installed and switched on; there is no lingering atmospheric legacy effect to overcome, no multi-decade cleanup process required, no equivalent of removing already-emitted carbon dioxide from the atmosphere. This single fact — that the problem, at the level of physical mechanism, essentially disappears the moment the responsible behavior stops — is the foundation for essentially everything else in this chapter.

Shielding and fixture design: the highest-leverage fix

The single most effective and most widely recommended intervention against terrestrial light pollution is full-cutoff fixture design: light fixtures engineered so that essentially all of their output is directed downward, toward the ground surface actually intended to be illuminated, with minimal or no light escaping directly upward or sideways above the horizontal. This is, in a meaningful sense, a solved engineering problem; full-cutoff and similarly shielded fixture designs have existed and been well understood for decades, and retrofitting an unshielded fixture with an appropriately designed shield, or replacing it entirely with a properly designed full-cutoff fixture, typically requires no exotic technology and, at scale, relatively modest incremental cost compared to a standard unshielded replacement of the same brightness and durability.

The International Dark-Sky Association and comparable organizations in multiple countries have developed detailed technical standards and fixture certification programs specifically to help municipalities, businesses, and individual property owners identify appropriately shielded lighting products, and a growing number of jurisdictions worldwide have adopted outdoor lighting ordinances requiring full-cutoff or comparably shielded fixtures for new public and, in some cases, private lighting installations, typically justified explicitly on energy-efficiency grounds (since properly shielded fixtures waste less light and therefore require less electricity to achieve the same useful ground-level illumination) as well as on dark-sky preservation grounds.

Color temperature: a subtler but significant factor

Chapter 6 noted a finding from recent sky-brightness research that deserves elaboration here: the ongoing global transition from older sodium-vapor street lighting to newer LED fixtures appears to have produced more sky brightening, as measured by both instruments and citizen-scientist naked-eye observation, than would have been predicted from the reduction in raw energy consumption LEDs typically achieve, a discrepancy substantially attributable to color temperature. Sodium-vapor lighting, the dominant technology for street and highway lighting through much of the twentieth century, produces a narrow-spectrum, orange-toned light that scatters somewhat less effectively through the atmosphere and to which the human eye's night-adapted (scotopic) vision is comparatively less sensitive. Many LED fixtures installed during the initial wave of municipal LED conversions beginning in the 2010s, by contrast, were selected primarily for cost and energy efficiency and defaulted to relatively cool, blue-enriched color temperatures, which both scatter more effectively through the atmosphere (contributing more skyglow per unit of raw light output) and register as more perceptually intense to human night vision, and, as discussed in Chapter 8, more effectively suppress human melatonin production as well.

The practical, and genuinely encouraging, implication of this finding is that warmer-color-temperature LED fixtures — increasingly available, and increasingly specified in updated municipal lighting standards and dark-sky-friendly lighting ordinances adopted since this research became widely known within the lighting-design and dark-sky advocacy communities — can substantially reduce both the skyglow and the human circadian impact of a given lighting installation without requiring any reduction in the fixture's useful ground-level brightness or any sacrifice of the substantial energy-efficiency gains LED technology offers over older lighting technology. This is very much the kind of intervention this chapter has emphasized throughout: a genuine improvement available through better-informed technical specification, not through simply asking people or municipalities to accept less light or spend more money.

Smart and adaptive lighting

A further, increasingly available category of intervention involves lighting controls — motion sensors, timers, and networked "smart" lighting systems capable of dimming or extinguishing fixtures during periods when their full brightness serves no practical purpose, such as overnight hours in commercial districts with minimal pedestrian or vehicle traffic, while restoring full brightness rapidly when motion or scheduled activity is detected. These systems address a specific and previously underappreciated feature of the traditional streetlight and security-lighting model: most such fixtures have historically operated at a single, constant brightness level throughout the entire night regardless of actual usage, a design choice driven largely by the simplicity and low cost of older lighting-control technology rather than any genuine need for constant maximum brightness at three in the morning in an empty parking lot. Adaptive lighting systems, now increasingly cost-competitive as networked LED control technology has matured and become cheaper, allow the same fixture to provide full safety-relevant brightness when and where it is actually needed, while substantially reducing both energy consumption and unnecessary skyglow contribution during the much larger fraction of the night when it is not.

Case study: Flagstaff, Arizona

Few examples demonstrate the practical achievability of comprehensive dark-sky policy better than Flagstaff, Arizona, home to Lowell Observatory (notably, the site of Pluto's discovery in 1930) and widely recognized as the first municipality in the world to adopt comprehensive outdoor lighting regulations specifically motivated by astronomical dark-sky preservation, beginning with an initial ordinance in 1958 and substantially expanded and strengthened multiple times in the decades since. Flagstaff's regulations address essentially every intervention category discussed in this chapter: mandated shielding requirements for both public and private outdoor lighting, restrictions on total permitted outdoor lighting levels for new commercial and residential development, specific limits on illuminated signage brightness and operating hours, and, in more recent updates, explicit color-temperature restrictions reflecting the LED-era research described above.

The city was formally recognized as the world's first International Dark Sky Community by the International Dark-Sky Association in 2001, a designation reflecting decades of sustained policy commitment rather than a single ordinance, and monitoring data collected at Lowell Observatory and by other researchers studying the city's sky brightness over the intervening decades has generally shown Flagstaff's sky brightness increasing far more slowly than comparably sized cities without equivalent lighting ordinances, despite substantial population and commercial growth in the city over the same period — clear evidence that dark-sky-friendly policy, sustained over a period of decades, can meaningfully decouple a growing city's lighting footprint from its actual sky-brightness trajectory, rather than the two being locked together as an unavoidable consequence of growth itself. Flagstaff's experience has subsequently informed dark-sky ordinance design in a considerable number of other municipalities worldwide, and its sustained, multi-decade track record makes it one of the strongest available pieces of evidence that comprehensive, well-designed lighting policy is not merely theoretically appealing but practically achievable and durable over the long term.

Satellite-side mitigation, briefly revisited

Chapters 11 and 12 already described the satellite-side mitigation efforts undertaken by at least some operators since roughly 2019 in direct response to astronomical community advocacy: darkening coatings and sun-shading design elements intended to reduce satellite reflectivity and therefore visible brightness, and coordination systems providing observatories with advance satellite position data to support scheduling and data-processing workarounds. These efforts are real, and in the specific cases where they have been implemented, have produced measurable brightness reductions for the satellites involved. They remain, however, voluntary commitments by individual operators rather than binding requirements applied consistently across the entire industry, and the policy gap identified in Chapter 13 — the absence of any international regulatory body with clear authority and mandate to weigh astronomical and general public interest in an undisturbed night sky against the commercial case for continued satellite deployment — remains fundamentally unaddressed as of this book's writing. Closing that gap, whether through new international agreements, expanded national regulatory mandates, or some other mechanism not yet clearly developed, represents the single largest remaining structural obstacle to a genuinely comprehensive response to the full scope of the problem this book has described, encompassing both its terrestrial and its orbital dimensions.

What individuals can actually do

Alongside policy-level interventions, a considerable amount of the terrestrial light pollution described throughout this book originates from choices made by individual homeowners, businesses, and property managers, over which national or even municipal policy has only partial influence, and which individual action can address directly and immediately. Simple, well-documented steps — installing shielded rather than unshielded outdoor fixtures, choosing warmer color-temperature bulbs for outdoor use, installing motion sensors or timers rather than leaving security and decorative lighting on continuously through the night, and simply using outdoor lighting at the minimum brightness and duration actually necessary for its intended purpose rather than defaulting to maximum brightness as a matter of habit — collectively represent a meaningful, achievable contribution to reducing a given neighborhood's or community's overall skyglow contribution, achievable without waiting for municipal policy change and, in most cases, without any sacrifice of genuine safety or convenience.

Cautious, earned optimism

This chapter closes, deliberately, on a note of cautious optimism, distinct from the more sobering tone of several preceding chapters, because the specific evidence assembled here supports it: light pollution, unlike some other major contemporary environmental challenges, has no lingering atmospheric legacy effect, responds essentially immediately to corrected behavior at the source, and has multiple decades of accumulated, real-world policy experience — exemplified by Flagstaff, but no longer unique to it — demonstrating that comprehensive, well-designed lighting policy can substantially slow or even reverse a community's sky-brightness trajectory without sacrificing safety, without abandoning modern lighting technology, and without imposing prohibitive cost.

The orbital half of this book's story remains considerably less resolved, both technically and, especially, in terms of the governance gap described in Chapter 13, and this chapter has not pretended otherwise. But even there, the fact that at least some satellite operators have engaged constructively with astronomical community concerns, and that measurable brightness reductions have already been achieved through voluntary design changes, suggests that meaningful mitigation, if pursued with sufficient institutional commitment and appropriately binding regulatory backing, is technically achievable there as well, even if the harder work of building an adequate international governance framework for a genuinely shared, borderless resource remains substantially undone.

The book's final section returns, one last time, to where it began: a beam of light atop a Nevada casino, and the question that beam has represented throughout this book's argument about what, exactly, humanity owes to the night sky it inherited.

Building codes and the "invisible" policy lever

One of the more effective but less publicly visible policy mechanisms for reducing terrestrial light pollution operates not through dedicated, freestanding dark-sky ordinances of the kind Flagstaff pioneered, but through incorporation of outdoor lighting standards into a jurisdiction's broader building and zoning codes — the same regulatory mechanism that governs everything from structural safety requirements to parking-space minimums for new commercial development. A growing number of jurisdictions, particularly in the years since the International Dark-Sky Association's model lighting ordinance became widely available as a template, have incorporated shielding, brightness, and color-temperature requirements directly into the standard permitting process for new construction and major renovation, meaning that compliance becomes a routine, largely automatic part of any new development's design and approval process, rather than requiring separate, freestanding advocacy or enforcement effort specifically focused on lighting.

This building-code approach has a specific practical advantage worth noting: it applies prospectively, to new construction and major renovations, without requiring the more politically and administratively difficult process of mandating retrofits to existing lighting installations, which represents, in most jurisdictions that have pursued comprehensive dark-sky policy, a slower, longer-term component of the overall lighting-improvement strategy, addressed gradually as existing fixtures reach the end of their operational life and are replaced in the ordinary course of maintenance, rather than through a disruptive, immediate mandatory replacement program. Over a sufficiently long time horizon — the multi-decade horizon over which Flagstaff's ordinance has operated, for instance — this gradual, replacement-driven approach can achieve very substantial cumulative improvement without ever having required the kind of immediate, costly, and politically difficult mandatory retrofit campaign that might otherwise generate significant opposition from affected property owners.

The limits of what policy alone can achieve

It would be an incomplete and overly optimistic account of this chapter's subject to suggest that policy interventions of the kind described above are sufficient, on their own, to fully resolve the terrestrial light pollution problem this book has documented at length, and it is worth closing this section of the chapter with a clear-eyed acknowledgment of the genuine remaining limits.

Enforcement of outdoor lighting ordinances, once adopted, requires ongoing municipal capacity and political will that not every jurisdiction sustains consistently over time, and a number of dark-sky advocates have noted, in various policy-evaluation contexts, that ordinances adopted with initial enthusiasm can sometimes suffer from inconsistent enforcement in later years, particularly in municipalities facing competing budgetary and administrative priorities. Commercial and advertising lighting, of the kind discussed in Chapter 4's treatment of neon signage and illuminated billboards, has historically proven more resistant to comprehensive regulation than residential and general commercial fixture shielding, both because illuminated signage frequently receives some degree of protection under commercial free-expression legal doctrines in some jurisdictions, and because the economic interests favoring maximal advertising visibility remain, as Chapter 4 discussed at length, directly and explicitly opposed to the brightness reduction that comprehensive dark-sky policy would otherwise favor.

And even the most comprehensive local or national terrestrial lighting policy, however well designed and consistently enforced, does nothing whatsoever to address the orbital half of this book's story — the satellite megaconstellation growth discussed in Chapters 11 through 14 — which, as those chapters emphasized at length, remains subject to a genuine, unresolved international governance gap that no amount of well-designed municipal or even national terrestrial lighting policy can, by itself, close.

A summary table of interventions and their maturity

It may help, at this point in the chapter, to summarize the range of interventions discussed across this book in terms of how mature and well-proven each approach currently is. Full-cutoff fixture shielding and warmer color-temperature specification represent mature, well-understood, relatively low-cost interventions with decades of accumulated real-world implementation experience and strong supporting evidence of effectiveness, exemplified most clearly by Flagstaff's sustained, multi-decade track record. Smart and adaptive lighting control systems represent a somewhat newer but rapidly maturing intervention category, increasingly cost-competitive and increasingly incorporated into updated municipal lighting standards, though with a shorter accumulated track record than fixture shielding alone. Satellite-side design mitigation (darkening coatings, sun shades) and coordination systems represent a genuinely new, actively evolving intervention category, with real but still-developing evidence of effectiveness and, critically, no binding regulatory requirement compelling universal adoption across the industry. And international governance mechanisms adequate to the full scope of the orbital light-pollution problem represent, as repeatedly emphasized throughout the second half of this book, the least mature and least developed intervention category of all, essentially still in the earliest stages of being conceived, let alone implemented, as of this book's writing.

This uneven state of development across intervention categories is itself an important part of this book's honest overall assessment: some parts of this problem are genuinely close to solved, at least in the specific jurisdictions that have chosen to adopt the available, well-proven solutions; other parts remain, as of this writing, substantially unresolved, and will likely require sustained attention, advocacy, and institutional development well beyond the publication of this book to adequately address.

Certification and consumer-facing labeling

A further, more recently developed mitigation mechanism worth noting is the emergence of formal dark-sky-friendly product certification for outdoor lighting fixtures themselves, administered by the International Dark-Sky Association and similar organizations, allowing manufacturers to have specific fixture models formally certified as meeting defined shielding, brightness, and color-temperature standards, and allowing consumers, municipal purchasing agents, and building developers to identify compliant products readily at the point of purchase, without needing specialized lighting-engineering expertise themselves to evaluate a given fixture's likely skyglow contribution. This certification and labeling infrastructure mirrors, in its basic structure, comparable consumer-facing certification systems developed for other environmental and efficiency goals — energy-efficiency labeling being the most obvious parallel — and has played a meaningful role in translating the more technical shielding and color-temperature research discussed earlier in this chapter into a form that ordinary purchasing decisions, whether by a municipal government procuring streetlights at scale or an individual homeowner selecting a porch fixture, can act on directly and with reasonable confidence, without needing to independently evaluate the underlying photometric specifications themselves.

A closing thought on the pace of possible change

It is worth ending this chapter by returning, briefly, to the pace-of-change theme that has recurred throughout this book. The historical chapters early in this book traced a century and a half of largely unexamined, accumulating lighting expansion; the more recent chapters have traced a rapid, still-unfolding decade or so of satellite megaconstellation growth. This chapter's discussion of solutions suggests that meaningful correction, at least on the terrestrial side of this book's argument, can in principle proceed considerably faster than the original problem accumulated, precisely because of the immediate reversibility this chapter opened with: a poorly designed streetlight replaced tonight stops contributing to skyglow immediately, without requiring anything like the multi-decade legacy cleanup process required by many other environmental harms. Flagstaff's more than six decades of sustained policy commitment demonstrate what is achievable with genuine, sustained institutional will; they also demonstrate that "immediately reversible in principle" and "actually reversed in practice" are two different things, requiring the kind of sustained, decades-long policy commitment this chapter has described, not a single ordinance passed and then forgotten. The orbital half of this book's story remains considerably further from any comparable resolution, but even there, this chapter has shown that meaningful, voluntary mitigation has already been achieved in specific cases, suggesting the technical and institutional capacity for further progress exists, if the political and regulatory will to fully mobilize it can be developed and sustained.

Retrofitting existing infrastructure at scale

Beyond new-construction building codes discussed earlier in this chapter, a number of municipalities have pursued dedicated, proactive retrofit programs specifically targeting existing streetlight inventories, often bundling dark-sky-motivated shielding and color-temperature improvements together with the broader LED conversion programs many cities have pursued primarily for energy-cost savings. This bundling strategy has proven a practically effective way to achieve dark-sky benefits alongside a program municipalities were often already planning to undertake for unrelated budgetary reasons, allowing the marginal additional cost of choosing dark-sky-compliant fixture specifications, rather than the cheapest available alternative, to be absorbed within a capital program whose primary justification and funding source rests on the energy-efficiency case alone.

Measuring policy success over realistic timeframes

A final, practically important point about the policy interventions catalogued throughout this chapter concerns the appropriate timeframe for evaluating their success. Flagstaff's own track record, discussed at length earlier in this chapter, took decades to fully demonstrate; a single year or even a single decade of a newly adopted ordinance is unlikely to produce dramatic, immediately visible improvement, particularly in a community starting from a heavily light-polluted baseline. This suggests that policymakers and advocates pursuing the interventions discussed in this chapter should set realistic, multi-decade expectations for measurable sky-brightness improvement, resisting any temptation to judge a newly adopted ordinance a failure simply because its full benefits have not yet become clearly visible within an unrealistically short evaluation window.

Utility-led rebate programs

A number of electric utilities, particularly in regions with strong existing energy-efficiency rebate programs, have extended those programs to specifically incentivize dark-sky-compliant outdoor fixture upgrades, recognizing the substantial overlap between energy efficiency and skyglow reduction discussed throughout this chapter, and providing a further, market-based funding mechanism that has helped accelerate adoption of the shielding and color-temperature improvements this chapter has identified as the most effective and best-evidenced interventions available.

International model ordinances and cross-border learning

The specific model lighting ordinance language developed by dark-sky advocacy organizations, referenced earlier in this chapter, has been adapted and translated for use well beyond its original country of origin, with municipalities in multiple countries drawing directly on the same underlying technical framework, adjusted for local legal and administrative context, illustrating a genuinely global diffusion of practical policy knowledge that mirrors, on the solutions side, the equally global scope of the underlying problem this book has documented throughout.

Epilogue — Look Up

Return, one final time, to midnight in Las Vegas.

The Luxor beam is still burning, exactly as it was on the first page of this book, xenon light gathered by mirrors and thrown straight up into the desert atmosphere with no destination and no purpose beyond its own existence. The insects are still rising into it in their hundreds of thousands. The bats are still arriving, night after night, to feed on a food source that exists purely because someone once decided a casino needed a more distinctive rooftop than its neighbors. Nothing about the scene has changed since the introduction, and that, in a sense, is exactly the point: the beam does not know, and does not care, that an entire book has just been written about what it represents. It simply keeps burning, as thousands of comparable fixtures across thousands of cities keep burning, because burning is what they were built to do, and no one with the authority to switch any of them off has yet been sufficiently persuaded that they should be.

But something else has changed since that opening scene, or should have: the reader's understanding of what that light actually costs, and where the story of that cost actually leads.

The progression, restated

This book has traced a specific arc, stated plainly in its introduction and, it is hoped, earned in the chapters since: we invented electric light, and used it so enthusiastically and so thoughtlessly that we brightened our own skies past the point where most of us can see the galaxy we live in. In response — though rarely framed this way at the time — we built telescopes of extraordinary size and sensitivity on the darkest, most remote mountaintops we could find, spending billions of dollars specifically to recover, through instruments, the view of the universe our own cities had taken from us with the naked eye. And now, in the middle of that immense and still-ongoing effort, we have begun filling the very medium those telescopes look through — the sky itself — with tens of thousands of our own bright, reflective, communicating machines, launched for reasons that have nothing to do with astronomy and everything to do with a genuinely different, but equally human, ambition: to connect every remaining unconnected person on the planet to the internet, as fast as commercially possible.

There is no villain in this story, and this book has tried, throughout, to resist the temptation to supply one. The engineers who built unshielded streetlights in 1905 were not trying to erase the Milky Way; they were trying to make city streets safer to walk at night, with the tools and knowledge available to them at the time. The casino executives who commissioned the Luxor beam were not trying to disrupt a desert ecosystem; they were trying to build a landmark distinctive enough to draw visitors past a hundred competing resorts. The companies launching satellite constellations by the thousand are not trying to make astronomy harder; they are trying to bring broadband internet to people who have never reliably had it, a goal whose value this book has never disputed. Every actor in this story has been pursuing a genuinely defensible, often genuinely valuable goal. The night sky's disappearance has been, almost entirely, a side effect — an externality, in the economist's term, that nobody explicitly chose, that nobody is fully responsible for in isolation, and that has therefore been remarkably difficult to address through the ordinary mechanisms by which societies usually correct course when a single identifiable actor causes a clearly attributable harm.

What Flagstaff and Paranal both teach

And yet — this book has also tried to show, honestly, that this diffuse, no-single-villain quality does not make the problem unsolvable. Flagstaff, Arizona proved, beginning in 1958 and sustained for more than six decades since, that a community can choose, deliberately and durably, to decouple its own growth from the brightening of its own sky, without sacrificing safety or economic vitality, simply by treating darkness as a resource worth the modest cost of protecting. Paranal, by contrast, is currently demonstrating the opposite lesson in real time: that even the most remote, most scientifically prized, most seemingly secure dark place on the planet is not automatically protected by its remoteness alone, and requires active, ongoing defense against development pressures that have nothing to do with astronomy and everything to do with entirely separate economic priorities that happen to intersect, geographically, with a resource nobody involved in that development was necessarily thinking about at all.

Both lessons matter, together. Darkness can be protected, deliberately and durably, when a community or an institution decides it is worth protecting and follows through with sustained policy commitment. And darkness is never protected by default, no matter how remote or seemingly secure a given location appears; it requires an active, ongoing decision, renewed again and again against the accumulating pressure of countless other actors pursuing entirely unrelated goals.

The choice this book has been building toward

Which returns us, finally, to the question this book has posed in various forms since its opening pages, stated here as plainly as it can be stated: is the night sky something we own — a resource available for whatever use any sufficiently motivated party can put it to, subject to no more restraint than local convenience and immediate commercial advantage happen to permit — or is it something we merely inherited, arriving to us fully formed, from a shared cosmic history vastly longer than the several hundred years humanity has spent, thus far, industrially rearranging its own visible surroundings — and something we are, in turn, obligated to hand on, in something reasonably close to the condition we received it, to whoever comes after us?

This book does not pretend to be neutral on that question, and by this final page, the reader has likely already guessed which answer it has been arguing for throughout. But the honest and important thing to say, in closing, is that the answer is not actually settled — not in law, not in international governance, not in the day-to-day decisions of the municipal planning boards, satellite-licensing regulators, and individual property owners whose accumulated choices will determine, in practice, what the night sky over the next several generations of human beings actually looks like. Every chapter of this book has been, in its way, an argument that the answer ought to tilt toward inheritance and obligation rather than ownership and convenience — but the argument alone does not decide anything. Only sustained, deliberate choice, at every level from a single homeowner selecting a shielded porch light to an international body eventually asserting some coordinated authority over orbital brightness, actually decides it, one accumulated decision at a time, exactly as it has been decided, so far, mostly by default and mostly in the other direction.

A closing invitation, not a conclusion

So this book will not end with a summary conclusion, because the story it has told is not actually concluded; it is, as of this writing, still very much in progress, its ending not yet determined. It will end instead with the same invitation offered in its opening pages, because that invitation is, in the end, the entire point of everything written since.

Go outside tonight. Find the darkest patch of sky you can reasonably reach — a backyard if that is what is available, a nearby park if it is a little better, a genuine dark-sky location if you are fortunate enough to have the time and means to travel there. Wait a few minutes for your eyes to adjust. And look up.

What you see, or fail to see, is not a fixed, permanent fact about the universe. It is a running tally of decisions — yours, your neighbors', your city's, a handful of satellite companies', and, ultimately, all of ours together — about how much of the sky we have decided, so far, that we are willing to give up. That tally is still being kept. It is not too late to change what it will eventually say.

Look up. And then decide what you want the next entry in that tally to be.

A final image

Picture, one more time, the two scenes with which this book began: the beam above the Luxor, and the astronomer waiting for photons older than the Earth itself. Hold them in mind together, side by side, for a moment, because that is really what this entire book has asked of its reader from the first page to this one – not to choose between them, not to condemn the beam or sanctify the telescope, but to see clearly that they exist in the same sky, are subject to the same physics, and are the product of the same species, arguing, in effect, with itself, about what that shared sky is actually for.

Both scenes are still unfolding tonight, as they were unfolding on the night this book opened, and as they will very likely still be unfolding on whatever night you happen to be reading these final pages. The beam has not been switched off. The satellites have not stopped launching. The astronomer is still waiting, patiently, for a signal that may or may not survive the journey down through an increasingly crowded sky. Nothing about the material facts of the story has been resolved by the writing of this book, and nothing was ever going to be. Books do not switch off beams or ground satellite launches. What they can do, at their best, is change what the people who read them notice, the next time they happen to look up.

So: look up. That is where this book started, and that is where it ends.

One last number worth remembering

If a single number from this book deserves to stay with the reader longest, it may be this one, restated here deliberately at the close: roughly a third of the human species, and by some measures four out of every five people in North America, can no longer see the galaxy they live in from the place they call home. That number was not true for any prior generation in the history of our species. It became true within living memory. And whether it becomes truer still, or begins, at last, to recede, is not a settled scientific fact about the universe. It is a policy choice, made one fixture, one ordinance, one satellite launch at a time.

www.ingramcontent.com/pod-product-compliance
Lightning Source LLC
LaVergne TN
LVHW050810061126
851857LV00012B/748